M. S. Dresselhaus R. Kalish

Ion Implantation in Diamond, Graphite and Related Materials

With 108 Figures

Springer-Verlag
Berlin Heidelberg New York
London Paris Tokyo
Hong Kong Barcelona
Budapest

Springer Series in ***Materials Science***

1 **Chemical Processing with Lasers**
By D. Bäuerle

2 **Laser-Beam Interactions with Materials**
Physical Principles and Applications
By M. von Allmen

3 **Laser Processing of Thin Films and Microstructures**
Oxidation, Deposition and Etching of Insulators
By I. W. Boyd

4 **Microclusters**
Editors: S. Sugano, Y. Nishina, and S. Ohnishi

5 **Graphite Fibers and Filaments**
By M. S. Dresselhaus, G. Dresselhaus, K. Sugihara, I. L. Spain, and H. A. Goldberg

6 **Elemental and Molecular Clusters**
Editors: G. Benedek, T. P. Martin, and G. Pacchioni

7 **Molecular Beam Epitaxy**
Fundamentals and Current Status
By M. A. Herman and H. Sitter

8 **Physical Chemistry of, in and on Silicon**
By G. F. Cerofolini and L. Meda

9 **Tritium and Helium-3 in Metals**
By R. Lässer

10 **Computer Simulation of Ion-Solid Interactions**
By W. Eckstein

11 **Mechanisms of High Temperature Superconductivity**
Editors: H. Kamimura and A. Oshiyama

12 **Dislocation Dynamics and Plasticity**
By T. Suzuki, S. Takeuchi, and H. Yoshinaga

13 **Semiconductor Silicon**
Materials Science and Technology
Editors: G. Harbeke and M. J. Schulz

14 **Graphite Intercalation Compounds I**
Structure and Dynamics
Editors: H. Zabel and S. A. Solin

15 **Crystal Chemistry of High-T_c Superconducting Copper Oxides**
By B. Raveau, C. Michel, M. Hervieu, and D. Groult

16 **Hydrogen in Semiconductors**
By S. J. Pearton, J. W. Corbett, and M. Stavola

17 **Ordering at Surfaces and Interfaces**
Editors: A. Yoshimori, T. Shinjo, and H. Watanabe

18 **Graphite Intercalation Compounds II**
Transport and Electronic Properties
Editors: H. Zabel and S. A. Solin

19 **Laser-Assisted Microtechnology**
By S. M. Metev and V. P. Veiko

20 **Microcluster Physics**
By S. Sugano

21 **The Metal-Hydrogen System**
Basic Bulk Properties
By Y. Fukai

22 **Ion Implantation in Diamond, Graphite and Related Materials**
By M. S. Dresselhaus and R. Kalish

Springer Series in **Materials Science** 22

Prof. M. S. Dresselhaus, Ph. D.
Department of Electrical Engineering and Computer Science
Massachusetts Institute of Technology
Cambridge, MA 02139, USA

Prof. R. Kalish, Ph. D.
Solid State Institute and Physics Department, Technion–Israel Institute of Technology
Haifa, 32000, Israel

ISBN 3-540-54956-0 Springer-Verlag Berlin Heidelberg New York
ISBN 0-387-54956-0 Springer-Verlag New York Berlin Heidelberg

Library of Congress Cataloging-in-Publication Data. Dresselhaus, M. S. Ion implantation in diamond, graphite and related materials / M. S. Dresselhaus, R. Kalish. p. cm. – (Springer series in materials science ; v. 22) Includes bibliographical references and index. ISBN 3-540-54956-0 – ISBN 0-387-54956-0 1. Ion implantation. 2. Diamond–Effect of radiation on. 3. Graphite–Effect of radiation on. I. Kalish, Rafael. II. Title. III. Series. Q702.7.I55D74 1992 620.1'98–dc20 92-11325

Printed in the United States of America

Typesetting: Camera ready by authors
54/3140-5 4 3 2 1 0 - Printed on acid-free paper

Preface

Carbon has always been a unique and intriguing material from a fundamental standpoint and, at the same time, a material with many technological uses. Carbon-based materials, diamond, graphite and their many derivatives, have attracted much attention in recent years for many reasons. Ion implantation, which has proven to be most useful in modifying the near-surface properties of many kinds of materials, in particular semiconductors, has also been applied to carbon-based materials. This has yielded, mainly in the last decade, many scientifically interesting and technologically important results. Reports on these studies have been published in a wide variety of journals and topical conferences, which often have little disciplinary overlap, and which often address very different audiences. The need for a review to cover in an integrated way the various diverse aspects of the field has become increasingly obvious. Such a review should allow the reader to get an overview of the research that has been done thus far, to gain an appreciation of the common features in the response of the various carbon allotropes to ion impact, and to become aware of current research opportunities and unresolved questions waiting to be addressed. Realizing this, and having ourselves both contributed to the field, we decided to write a review paper summarizing the experimental and theoretical status of ion-implantation into diamond, graphite and related materials. It did not take long, however, to realize that the topic is too broad to be covered in just a review article; hence it was decided to extend the manuscript into a comprehensive tutorial book.

The present book surveys the current status of research involving ion implantation into diamond and its derivatives, diamond-like a-C:H films and the very recent synthetic polycrystalline diamond films, as well as implantation into graphite, in its many forms, glassy carbon, carbon fibers and disordered carbon. For background, a short presentation is given on the structure and properties of these materials prior to implantation. A tutorial section is also devoted to the general description of ion implantation and the physics of ion-solid interactions, which describe the damage that always accompanies ion implantation. A review is included of the different experimental techniques employed to assess the consequences of the implantation, and of post-implantation thermal treatments. With this background in hand, the reader, who might be a new graduate student entering the field, a researcher with limited knowledge of the field, or a researcher active in studying any one of the carbon-based materials covered in the book, can now appreciate in a broader sense the results of ion implantation into carbon-based materials, which constitute the major part of the book. The book is written in a form and style which should make it comprehensible and useful to the wide spectrum of researchers interested in the topic, whether they are physicists, chemists, materials scientists and engineers, electrical engineers or mechanical engineers.

Because of the extensive current research activity in the field and the diverse journals in which this research is published, we found it extremely difficult to keep track of the many recent publications. However, we hope to have covered the majority of important results up to the end of 1990. Also included are a few of the most significant 1991 publications. Altogether there are over 300 references in the book, mostly from the last 10 years. Nevertheless, some important publications have no doubt been overlooked.

We hope that this book will offer a useful overview of the effect of ion implantation on the many, very different, carbon-based materials. The inter-comparison may thus allow the reader to see the features which are common to all the materials discussed in this book, and possibly trigger new research on both basic and applied aspects in this family of materials of great technological importance and of much scientific interest.

We wish to thank Dr. G. Dresselhaus, Dr. G. Braunstein, Professor M. Endo, Dr. S. Prawer, and Dr. B Elman for numerous enlightening discussions related to this review. One of us (MSD) gratefully acknowledges support from NSF grant DMR88-19896.

Cambridge
Haifa
February 1992

M.S. Dresselhaus
R. Kalish

Contents

1. Introduction

Carbon is unique in its properties, being the basis for a large group of materials very important to civilization as we know it. Carbon is basic to organic chemistry, and hence is the corner stone to all living organisms and to world energy supplies; it forms the basis for most polymers and the industries based on them.

From a materials science standpoint carbon is unique in supporting both sp^2 bonding, giving rise to graphite, a highly anisotropic electrical conductor, and sp^3 bonding, giving rise to diamond, a wide gap semiconductor (insulator). Though both materials are rather inert under ambient temperatures and pressures, they can transform into one another when exposed to special conditions.

Ion implantation is commonly used as the method of choice to modify many of the near-surface properties of materials. The major applications of ion-implantation are in the field of micro-electronics, where ion implantation has in many cases replaced diffusion as a means of introducing dopant atoms into semiconductors, especially for the group IV isoelectronic, III–V and II–VI compound semiconductors. Such doping applications also apply in principle to diamond. However, as will be shown below, the doping of diamond by ion implantation is much more complicated than the doping of other more common semiconductors. Ion implantation is used in fields other than semiconductors to improve the mechanical properties of materials or in synthesizing new phases. These uses have also been applied to graphite, diamond, or other carbon-based materials. In addition, ion implantation has been used as a method of introducing controlled amounts of lattice damage. In the case of graphite this has allowed a detailed study of the graphitization process, as the lattice damage is subsequently annealed. In the case of diamond, implantation-induced damage has allowed study of the transformation of metastable crystalline diamond to an amorphous structure and to the most stable form of carbon, namely graphite.

The study of the effects that ion-bombardment has on carbon-based materials is also of technological importance as they may be encountered in various technologies leading to either desirable or undesirable alterations in materials properties. For example, diamond devices (when they become available) may be suitable for special applications in the outer-space environment where they may be subjected to ion impact which may degrade their performance. Diamond-like coatings are now being used as protective layers for a variety of delicate components, and the understanding of their response to ion-irradiation is of technological importance; graphite is used as the material of choice for limiters in Tokomaks, and diamond-like amorphous hydrogenated carbon (a–

C:H) coatings are used as protective layers on the internal parts of Tokomaks, both operating under conditions in which they are exposed to severe ion bombardment. While ion implantation tends to degrade electrical performance, it often improves the mechanical properties of the affected near-surface regions; ion implantation is therefore often used as a method of choice for improving the near-surface mechanical properties of a variety of carbon-based materials. Thus, the understanding of the electrical, mechanical and other effects that ion implantation have on these materials is of great importance for their optimum utilization.

This book focuses on progress that has been made in the use of ion implantation to modify the properties of diamond, graphite, diamond-like films, and other carbon-related materials, and to deepen our understanding of their physical properties following ion implantation.

For background, the structure and properties of diamond, graphite and related carbon materials are very briefly reviewed in Chapter 2, with special emphasis given to the relation between the structure and properties of these materials and the sp^2 versus sp^3 bonding encountered in them. Then the ion implantation technique is reviewed in Chapter 3 with particular regard to the special properties of carbon materials as targets for the implanted ions. The major experimental techniques used to characterize ion-implanted diamond, graphite and related materials are discussed in Chapters 4 and 5. With this background in place, the progress made in recent years in studying ion implantation in diamond, graphite and other carbon-related materials is reviewed in Chapters 6, 7, 8 and 9. In the concluding remarks (Chapter 10), the similarities and differences between the various carbon-based ion-implanted materials reviewed in this book are summarized, and some possible future trends are outlined.

2. Carbon Materials: Graphite, Diamond and Others

Carbon materials are unique in many ways. One distinction relates to the various allotropic forms these materials can assume. Under ambient conditions, the graphite phase with strong in-plane trigonal bonding is the stable phase, as indicated by the phase diagram of Fig. 2.1 [2.1, 2]. Under the application of high pressure and high temperature (which are somewhat reduced when catalyst particles like iron or nickel are used), transformation to the diamond structure takes place. Once the pressure is released, diamond remains essentially stable under ambient conditions although, in principle, it will very slowly transform to the thermodynamical stable form of solid carbon which is graphite. However, when exposed to various perturbations, diamond will transform back to the

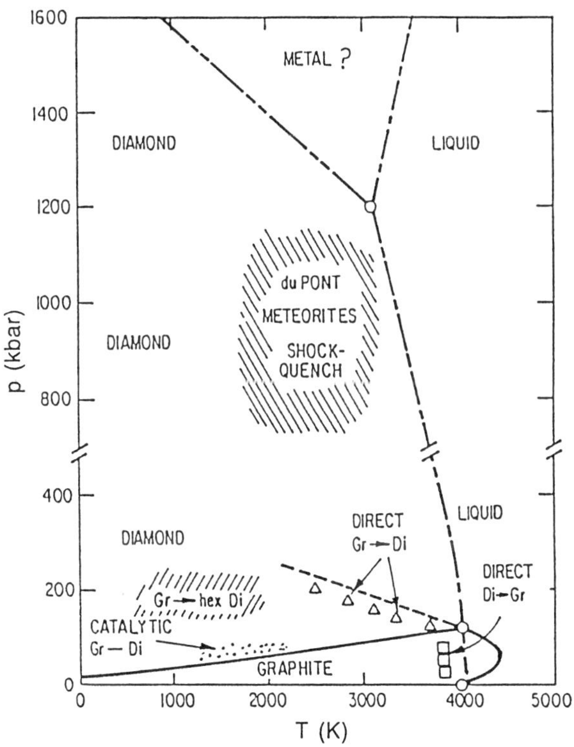

Fig. 2.1. One version of the phase diagram of carbon suggested by *Bundy* [2.2]

Table 2.1. Properties of graphite and diamond

	Graphite[a]		Diamond
Lattice structure	Hexagonal		Cubic
Space group	$P6_3/mmc - D_{6h}^4$		$Fd3m - O_h^7$
Atom locations in unit cell	(000), $(\frac{1}{3}\frac{2}{3}0)$,	$(00\frac{1}{2})$ $(\frac{1}{3}\frac{2}{3}\frac{1}{2})$	(000), $(\frac{1}{2}\frac{1}{2}0)$ $(0\,\frac{1}{2}\frac{1}{2})$, $(\frac{1}{2}0\frac{1}{2})$ $(\frac{1}{4}\frac{1}{4}\frac{1}{4})$, $(\frac{3}{4}\frac{3}{4}\frac{1}{4})$ $(\frac{1}{4}\frac{3}{4}\frac{3}{4})$, $(\frac{3}{4}\frac{1}{4}\frac{3}{4})$
Lattice constant (RT) [Å]	2.462	6.708	3.567
Atomic Density [cm^{-3}]	1.14×10^{23}		1.77×10^{23}
Specific gravity [g/cm^3]	2.26		3.515
Thermal conductivity (RT) [W/cm·K][b]	30	0.06	~25
Debye temperature [K]	2500[c]	950[c]	1860
Bulk modulus [$N{\cdot}m^{-2}$]	–		4–5.5×10^{11}
Elastic moduli [GPa]	1060[d]	36.5[d]	–
Band gap [eV]	–0.04		5.47
Electron mobility (RT) [$cm^2/(V\ s)$]	20000	100	1800
Hole mobility (RT) [$cm^2/(V\ s)$]	15000	90	1500
Dielectric constant (RT, low freq.)	3.0	5.0	5.58
Breakdown field [V/cm]	–	–	10^7 (highest)
Refractive index (visible)	–	–	2.4
Melting point [K]	4200		4500
Thermal expansion coefficient at RT [/K]	-1×10^{-6}	$+29 \times 10^6$	$\sim1\times10^{-6}$
Velocity of sound [cm/s]	$\sim 2.63 \times 10^5$	$\sim 1 \times 10^5$	$\sim 1.96 \times 10^5$
Raman frequency [cm^{-1}]	1582		1332

[a] Anisotropic quantities are given first as the in-plane (a-axis) value, then the c-axis value, where the a-axis value refers to the value in the ab plane.
[b] Highest reported thermal conductivity values are listed.
[c] Ref. [2.6].
[d] In-plane value is C_{11} and c-axis value is C_{33}.

equilibrium graphite phase. In this phase, the structure is highly anisotropic, exhibiting, for example, metallic behavior in the basal (ab) plane and poor electrical conductivity along the c-axis [2.3]. In contrast, diamond is an isotropic cubic wide gap semiconductor [2.4]. In terms of mechanical properties, graphite is the stiffest material in nature (has the highest in-plane elastic modulus), while diamond is the hardest (least deformable) material. Of all materials, diamond along with graphite (in-plane) exhibit the highest thermal conductivity and the highest melting point [2.5]. Diamond also has the highest atomic density of any solid (Table 2.1).

Atomic carbon has an atomic number of 6 and a $1s^22s^22p^2$ electronic ground state configuration. In the graphite structure, strong in-plane bonds are formed between a carbon atom and its three nearest-neighbors from $2s$, $2p_x$ and $2p_y$ orbitals; this bonding arrangement is denoted by sp^2. The remaining electron with a p_z orbital provides only weak interplanar bonding, but is responsible

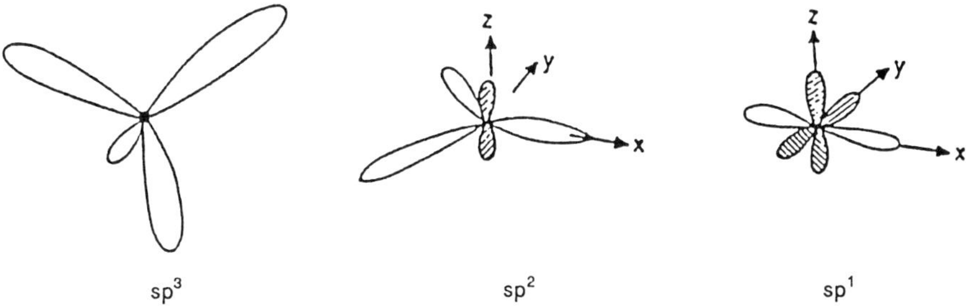

Fig. 2.2. Schematic presentation of sp^3, sp^2 and sp^1 hybridizations [2.7, 8]. The open loops denote strong bonds and the shaded loops denote weak bonds

for the semimetallic electronic behavior in graphite. In contrast, the carbon atoms in the diamond structure are tetrahedrally bonded to their four nearest neighbors using $2s$, $2p_x$, $2p_y$ and $2p_z$ orbitals in an sp^3 configuration (Fig. 2.2). The difference in the structural arrangement of these allotropic forms gives rise to the wide differences in their physical properties. It is this difference in structure and properties that is the focus of Chap. 2.

This chapter also focuses on the connection between graphite and diamond, which has over the years attracted a good deal of attention. Recently it has been found from theoretical calculations that the sp^3 diamond bonding should become more stable than the sp^2 graphite bonding when the vacancy density is above 8% [2.9]. Such vacancy densities can perhaps be achieved by ion implantation, so that this prediction can be experimentally verified. If indeed found to be correct, this finding could be of great interest for the low temperature synthesis of diamond films.

2.1 Structure and Materials

2.1.1 Graphite

The ideal crystal structure of graphite (Fig. 2.3) consists of layers in which the carbon atoms are arranged in an open network, such that the A and A' atoms on consecutive layers are on top of one another, but the B atoms in one plane are over the unoccupied centers of the adjacent layers, and similarly for the B' atoms [2.10]. This gives rise to an ABAB planar stacking arrangement shown in Fig. 2.3, with an in-plane nearest-neighbor distance of 1.421Å, an in-plane lattice constant of 2.462Å, a c-axis lattice constant of 6.708Å and an interplanar distance of 3.3539Å (Table 2.1). This structure is described by the $D_{6h}^4(P6_3/mmc)$ space group and has 4 atoms per unit cell (see Fig. 2.3).

When graphite is under deformation, a second layer stacking sequence, the rhombohedral ABCABC sequence, is introduced [2.3]. Disorder tends to have little effect on the in-plane lattice constant, largely because the in-plane C–C bond is very strong and the C–C spacing is very small; disorder does however

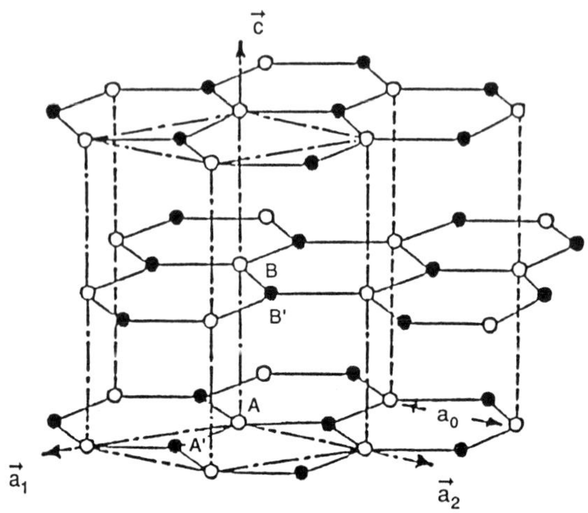

Fig. 2.3. The crystal structure of hexagonal single crystal graphite with D_{6h}^4 $P6_3/mmc$ symmetry, in which planes of carbon hexagons are stacked in an ...ABAB... sequence [2.10]. The A and A′ carbon sites are denoted by open circles and the B and B′ sites by black circles. The in-plane lattice constant is denoted by a_0, and the vectors of the unit cell in the directions $\mathbf{a}_1$, $\mathbf{a}_2$ and $\mathbf{c}$ are indicated

have a significant effect on the interplanar spacing because of the weak interplanar bonding. One consequence of this is that impurity species are unlikely to enter the in-plane lattice sites substitutionally, but rather will occupy some interstitial position between the layer planes.

Weak disorder results in stacking faults giving rise to a small increase in the interlayer distance until a value of about 3.440Å is reached, at which distance the stacking of the individual carbon layers (called graphene layers) becomes uncorrelated; the resulting two-dimensional (2D) structure of uncorrelated graphene layers is called turbostratic graphite [2.11]. As is shown below, the electronic structure of the turbostratic graphite, a zero gap semiconductor, is qualitatively different from that of ideal graphite, a semimetal with a small band overlap. Likewise, as the disorder is increased, so is the interplanar spacing which further modifies the electronic properties.

Graphite single crystals tend to be very small in size (on a millimeter scale) and very thin (on a 0.1mm scale), and for this reason are not useful for many scientific studies and practical applications. For this reason a variety of graphite-related materials are used for scientific studies and for practical applications. A very brief review of these materials is given below.

2.1.2 Graphite-Related Materials

The comparison material in all graphite studies is the natural single crystal graphite flakes that are found in several locations around the world, especially in Madagascar, the USSR, and the Ticonderoga area of New York State in the

US. The natural crystal flakes can sometimes be as large as several mm in the basal plane, and are typically much less than 0.1 mm in thickness. Natural graphite flakes usually contain several twinning planes, so that careful selection of flakes is necessary for detailed structural studies. The natural flakes also contain impurities which are chemically removed in a purification step, by boiling the flakes in concentrated HF and/or heating the flakes (to $\sim$ 2000°C) in flowing fluorine gas. Even after such chemical purification, metallic impurities (especially transition metals such as Fe) remain in the material at the ppm level [2.12]. Because of their small size, very few ion implantation studies have been carried out on natural graphite crystals.

A synthetic single crystal graphite called kish graphite is commonly used in scientific investigations. Kish-graphite crystals form on the surface of high carbon content iron melts, and are harvested as crystals from such solutions [2.13]. The as-grown kish-graphite flakes are subsequently purified in much the same way as natural-graphite flakes [2.12]. There are several reasons for using kish graphite in scientific studies. Firstly, some kish-graphite flakes can be found with relatively large sizes in the basal plane, relative to natural-graphite flakes. Secondly, very small kish graphite flakes ($\leq$ 1mm in size) without twin boundaries and with a relatively low density of defects can be found. Thus kish graphite is often used as a material of choice for sophisticated structural studies. Thirdly, the quality of selected kish-graphite flakes can be very high (residual resistivity ratios exceeding 100 have been reported [2.14]), relative to selected natural single crystal graphite flakes. A few significant ion implantation studies have been carried out on kish-graphite flakes.

However, most of the important ion implantation studies on graphite have been carried out on highly oriented pyrolytic graphite (HOPG), which is prepared by pyrolysis of hydrocarbons at temperatures above 2000°C and is subsequently heat treated to higher temperatures [2.15]. When stress annealed above 3300°C, the HOPG exhibits electronic, transport, thermal and mechanical properties close to those of single crystal graphite, showing a very high degree of c-axis alignment. For the high temperature, stress-annealed HOPG, the crystalline order extends to about 1μm within the basal plane, and to about 0.1μm along the c-direction. For all HOPG material however, there is no long-range in-plane a-axis alignment, and the a-axes of adjacent crystallites are randomly arranged [2.15].

The degree of structural order and c-axis alignment can be varied by control of the major processing parameters: heat treatment temperature T_{HT}, residence time at T_{HT}, and applied stress during heat treatment [2.15]. Turbostratic pyrolytic graphites are obtained for $T_{HT} < 2300$°C, and higher T_{HT} values are needed to establish 3D ordering. As stated above, most ion implantation studies on graphite have been carried out on the class of materials denoted by HOPG, but only a subset of HOPG materials have in fact been used for most of the ion implantation studies that have been carried out. This subset of HOPG materials is prepared by stress-annealing at $T_{HT} \geq 3200$°C to achieve a small spread in c-axis orientations (less than 0.5°) and a large in-plane crystallite size (in excess of 1 μm) [2.15]. Use of the most highly ordered HOPG materials allow ion beam channeling studies to be carried out (see Sect. 4.7). The rich materials science literature on the graphitization of HOPG and other carbons is extensively used in the interpretation of ion implantation and subsequent annealing studies in graphite.

Recently new precursor materials, such as polyimide (PI) and polyoxadiazole (POD) resins [2.16-20] have been used to synthesize graphite films, and these films show a high degree of 3D structural ordering when heat treated to $T_{HT} > 2800°$C. The quality of these films, especially those based on the KAPTON and NOVAX (polyimide) precursors, is rapidly improving and they may soon become materials of choice for specific ion implantation and annealing studies on graphite. To date, no ion implantation studies on these graphite films have been reported. One potential application of these graphite films may be as a high thermal conductivity electrical conductor in contrast to diamond which has an equally high thermal conductivity but is electrically insulating; for such an application, ion implantation with suitable masks could be used to define the conduction path.

In addition to highly crystalline graphite, several less ordered phases such as carbon fibers and glassy carbon are of great interest for practical applications (Sects. 2.1.3 and 2.1.4).

2.1.3 Carbon Fibers

Carbon fibers represent another class of graphite-related materials on which ion implantation work has been done. Despite the many precursors that can be used to synthesize carbon fibers, each having different cross-sectional morphologies (Fig. 2.4), the preferred orientation of the fiber axes for all carbon fibers is close to an a-axis of a graphene layer or fragment, thereby accounting for the high mechanical strength of these fibers [2.11]. Referring to the various morphologies in Fig. 2.4, the as-prepared vapor grown fibers have an onion skin morphology (Fig. 2.4a), which after heat treatment to about 3000°C forms facets (Fig. 2.4b); of all carbon fibers, these faceted fibers are closest to crystalline graphite in both crystal structure and properties. The commercially available mesophase pitch fibers with either the radial morphology (Fig. 2.4c) or the "PAN-AM" morphology (Fig. 2.4d) are exploited for their extremely high bulk modulus, while the commercial PAN (polyacrylonitrile) fibers with circumferential texture are widely used for their high strength [2.11]. Typical fiber diameters are $\sim 10\mu$m. The various morphologies shown in Fig. 2.4 exploit the highly anisotropic layered structure that is essentially unique to graphite. The small diameters of very thin free-standing vapor grown carbon fibers (in this case $< 1\mu$m diameter) have been exploited in controlled high resolution TEM studies of the implantation-induced disorder in highly graphitic fibers, and in studying the subsequent annealing of this disorder [2.21] (Sects. 6.1 and 6.2).

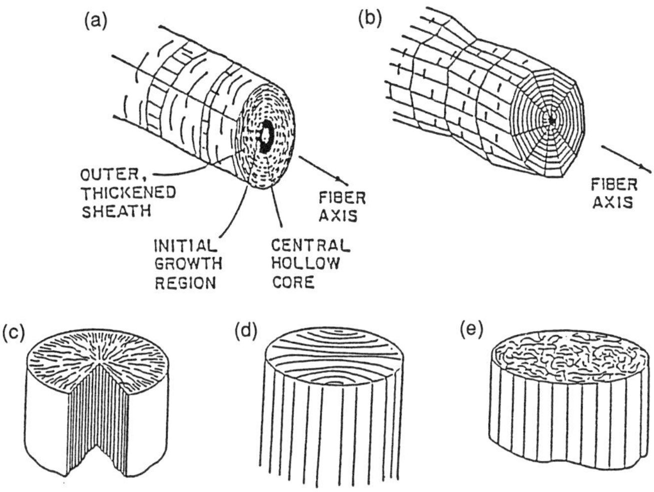

Fig. 2.4a-e. Sketch illustrating the morphology of vapor grown carbon fibers (VGCF): (**a**) as-deposited at 1100°C [2.11], (**b**) after heat treatment to 3000°C [2.11]. The morphologies for commercial mesophase-pitch fibers are shown in (**c**) for a "PAC-man" section with a radial arrangement of the straight graphene ribbons, and (**d**) for a "PAN-AM" section showing a transverse alignment with nearly parallel graphene planes, and (**e**) for a PAN fiber, with a circumferential arrangement of ribbons in the sheath region, and random in the core

2.1.4 Glassy Carbon

Glassy carbon (GC), another carbon material which has been used for ion implantation is manufactured as a commercial product by slow, controlled degradation of certain polymers at temperatures typically on the order of 900–1000° C [2.22]. This material is glass-like, granular, moderately hard, can be easily polished, is thermally conducting, impermeable, biocompatible, and stable at high temperatures. The apparent density of GC ranges from 1.46–1.50 g/cm^3 irrespective of heat treatment temperature, implying the existence of pores in the matrix. According to the *Jenkins* and *Kawamura* model [2.22], the microstructure of GC consists of a tangle of graphite-like ribbons or microfibrils, about 100 Å long and 30 Å in cross section (Fig. 2.5), and resembles the polymer chain configuration from which the GC has been derived. Because of the tangled ribbon microstructure, *Jenkins* and *Kawamura* have argued that glassy carbon does not fully graphitize, even when heat treated above 3000° C and for this reason glassy carbon is used as a prototype hard (non-graphitizable) carbon [2.23]. X-ray diffraction studies of the radial distribution function show that the carbon atoms are well ordered in the honeycomb in-plane structure of the graphene layers, but that the 3D registry between the graphene layers is poor, so that the ribbons form a turbostratic structure, typical of hard carbons [2.24, 25]. Recent structural studies [2.26] have shown the pores to form closed

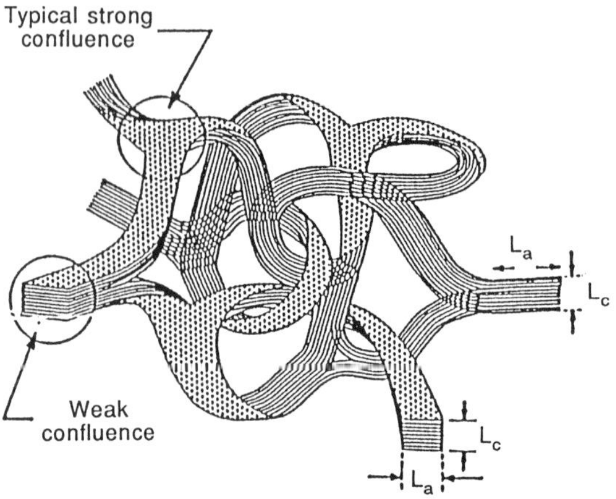

Fig. 2.5. Tangled structure proposed for many polymer derived graphitic carbons including glassy carbon. Note the presence of pores (hence a low mass density) which are likely sources of mechanical weaknesses [2.22]

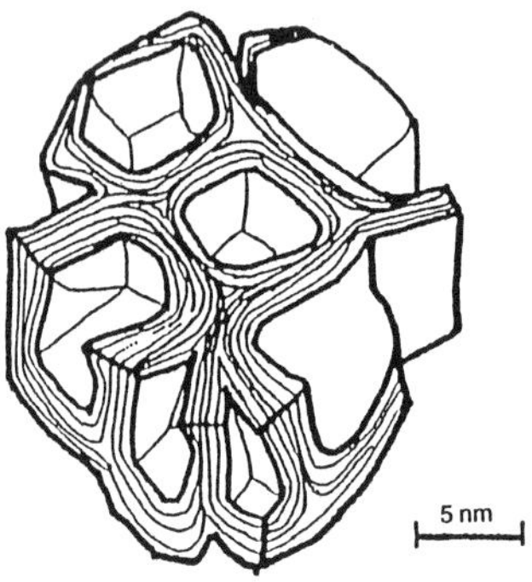

Fig. 2.6. Schematic diagram for the microstructure of the closed pore structure model for glassy carbon and other hard carbons [2.27, 28]

structures according to a model proposed by *Shiraishi* (Fig. 2.6) [2.27, 28] rather than an open interconnected network of ribbons [2.22]. The main arguments in support of the *Shiraishi* model for the closed pore network [2.26] are the microstructure and granular texture observed by SEM, and the good agreement between the measured apparent density and the structural determination of the apparent density based on measurements of the average grain size, the average interplanar distance, the average interplanar separation, and the average c-axis crystallite size L_c implied by the X-ray $(00l)$ linewidths. According to this model the average thickness of the pore walls (Fig. 2.6) is $L_c/2$. Since glassy carbon is a disordered graphite-like material that has been widely studied by many techniques (including TEM [2.29], SEE [2.30], and magnetoresistance [2.26]), it is not surprising that several groups have used glassy carbon for ion implantation studies (see Sect. 7.1).

2.1.5 Graphite Intercalation Compounds

Graphite intercalation compounds (GICs) are formed by the insertion of layers of guest species between the layers of the graphite host material [2.31, 32], as shown schematically in Fig. 2.7. The guest species may be either atomic or molecular. Because of the weak interlayer forces associated with the sp^2 bonding in graphite, the anisotropic layered graphite-based GICs can be synthesized. In diamond, on the other hand, the sp^3 bonding (Fig. 2.2) does not permit insertion of layers of guest species, and does not support intercalation. In the so-called donor graphite intercalation compounds, mobile electrons are transferred from the donor intercalate species (such as a layer of the alkali metal potassium) into the graphite layers, thereby raising the Fermi level E_F and increasing the electron concentration by two or three orders of magnitude, while leaving the intercalate layer positively charged with low mobility carriers. Conversely, for acceptor GICs, holes are transferred from the intercalate species (which is usually molecular) into the graphite layers, thereby lowering the Fermi level E_F. Because of the attractive in-plane interaction between the intercalate atoms or molecules, and the repulsive interplane interaction resulting from the intercalation-induced lattice strain, the intercalate layers form an ordered superlattice structure, interleaved with the graphite layers, through a

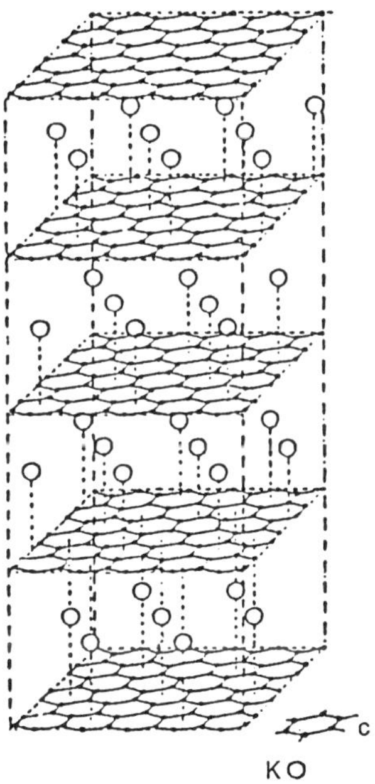

Fig. 2.7. Schematic model for a graphite intercalation compound showing the stacking of graphite layers (networks of small solid balls) and of intercalate (e.g., potassium) layers (networks of large hollow balls). For this stage 1 compound each carbon layer is separated by an intercalate layer [2.33]

phenomenon called staging [2.31, 32]. A GIC of stage n has isolated intercalate layers separated from one another by n graphite layers. Thus, a stage 4 K–GIC (see Sect. 4.9), will have each potassium layer separated by 4 graphite layers forming a unit cell of length $I_c = d_s + (n+1)c_0$ where $d_s = 5.35$Å is the thickness of a sandwich formed by two graphite layers between which the potassium layer is sandwiched and $c_0 = 3.35$Å is the interlayer distance between graphite layers. For an acceptor $SbCl_5$–GIC (see Sect. 4.8), the intercalate unit consists of a Cl–Sb–Cl trilayer with $d_s = 9.42$Å so that $I_c = 12.77$Å for a stage 2 $SbCl_5$–GIC. For the donor KHg–GIC (Sect. 4.8), the intercalate unit consists of a K–Hg–K trilayer and $d_s = 10.22$Å. Ion implantation has been used as a means of enhancing intercalation (Sect. 6.7) in these graphite-based materials.

2.1.6 Diamond

The diamond structure is probably the most important and most thoroughly investigated of all crystallographic structures. Silicon and germanium, the most commonly used elemental semiconductors, also exhibit the same "diamond" structure, while other important group III–V or II–VI compound semiconductors (such as GaAs or CdTe) crystallize in the closely related zincblende structure, the only difference between this and the diamond structure being that the two constituent atomic species of the zincblende structure occupy alternate sites in the diamond structure [2.10].

The ideal diamond structure shown in Fig. 2.8 has the characteristic property that every carbon atom is surrounded by four other carbon atoms at the corners of a regular tetrahedron with a cube edge length of $a_0 = 3.567$Å, and this carbon atom is bonded to these neighbors by strong covalent sp^3 bonds. The diamond structure is therefore cubic and can be viewed (Fig. 2.8) as two interpenetrating FCC structures displaced by $(1/4, 1/4, 1/4)a_0$ along the body diagonal. The nearest-neighbor carbon-carbon distance is 1.544Å, nearly 10% larger than in graphite, yet the density of diamond (1.77×10^{23} cm^{-3}) is 56% higher than in graphite, due to the high anisotropy of the graphite structure (Table 2.1). Also listed in Table 2.1 are the positions of the carbon atoms defining the basis vectors of the two FCC sublattices in the diamond structure. The diamond crystal, in contrast to graphite, is highly symmetric with a cubic space group $Fd3m - O_h^7$. Furthermore, diamond cleaves along {111} planes, while

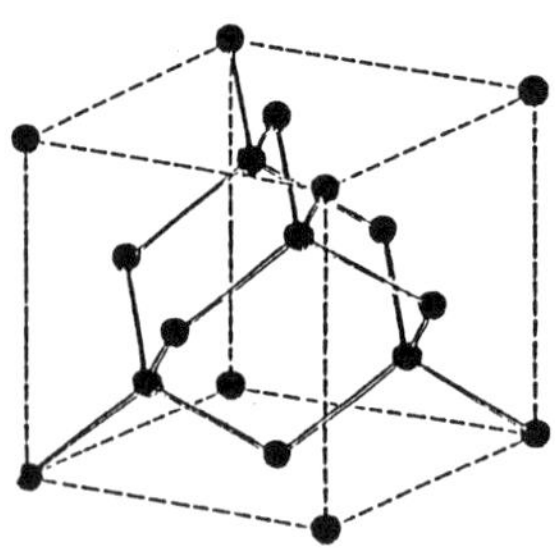

Fig. 2.8. The ideal diamond structure [2.10]

graphite cleaves along {001} planes [2.3]. There also exists a hexagonal form of diamond with space group D_{6h}^4 or $P6_3/mmc$, the same as for graphite, but with different site locations. The packing in hexagonal diamond is similar to that of cubic diamond, except for a shift of one of the carbon layers along the [111] planes [2.34]. Since no evidence for hexagonal diamond has been found in ion implanted diamond, further discussion of hexagonal diamond is not included.

Impurities in diamond are very important because of the changes they induce in the properties of diamond; these modified properties find applications mainly in industrial processes. The best natural diamonds contain impurities with concentrations of ~1 part in 10^5. Very few chemical species (e.g., B, N) can enter the diamond lattice substitutionally and, even when this is possible, the concentration of such substitutional impurities is very low (less than 1 part in 10^4). This situation is similar to graphite where substitutional impurities are limited to B; however, substitutional impurities are relatively less important in modifying the properties of graphite, as compared to diamond.

Diamonds have been historically classified according to their optical absorption properties, which are determined by impurities. The so-called type Ia diamonds exhibit strong absorption in the infrared. This is caused by fairly substantial amounts (up to 0.1%) of nitrogen, inhomogeneously distributed in the crystal and mainly concentrated in small agglomerates. Most natural diamonds belong to this group. Type Ib diamonds contain nitrogen as substitutional impurities. While the type Ib diamonds are rarely formed in nature, most synthetic diamonds belong to this group. Diamonds with the highest purity are type IIa, and these exhibit the intrinsic semiconducting properties of diamond with a wide band gap of 5.47 eV. Type IIb diamonds are naturally boron doped, and show *p*-type conductivity due to an acceptor level introduced by the substitutional boron located ~0.35 eV above the valence band.

Synthetic diamonds grown under conditions of high temperature and pressure have the same structure and defect types as natural diamonds.

2.1.7 CVD Diamond Films

In the last few years an important scientific and technological breakthrough occurred with the discovery that diamond thin films can be successfully grown by a large variety of chemical and physical vapor deposition techniques [2.35-39] . The feature common to all these methods is that cracked hydrocarbon radicals impinge upon a hot (~ 900°C), usually pre-stretched, surface in the presence of atomic hydrogen. Since this procedure enhances the formation of sp^3 over sp^2 bonds, diamonds can be grown. Diamond films have been prepared on a variety of substrates, including Si, quartz, Ni, and W. The films grown by these techniques are usually polycrystalline, consisting of agglomerations of randomly oriented, small diamond crystallites (several microns in size), and the films thus tend to have very rough surface morphologies. The structure of polycrystalline diamond films and their properties have recently been studied by *Narayan* [2.40] who has exposed the films to a series of microscopic evaluations.

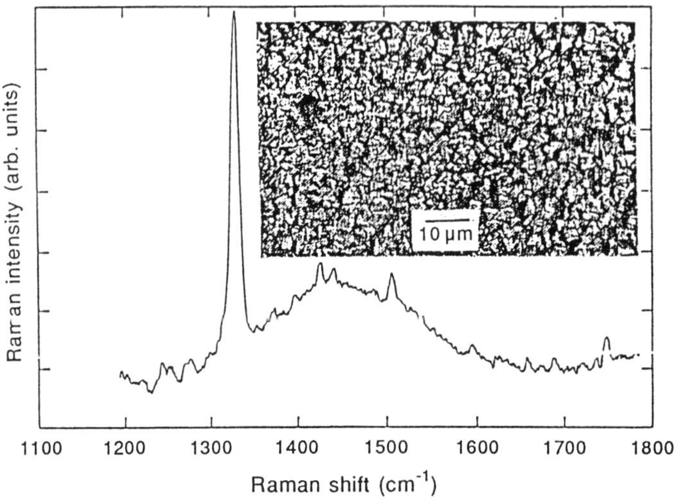

Fig. 2.9. Raman spectrum of the CVD diamond film grown on a fused quartz substrate. *Inset*: optical micrograph displaying the film morphology [2.41]

These included: scanning electron microscopy (SEM), which showed the gross structure of the films and exhibited the faceted nature of each grain; transmission electron microscopy (TEM), which revealed the defects in the material, the defects being mostly dislocations; and high resolution transmission electron microscopy (HRTEM), which showed the presence of twins in the material grown. Raman spectroscopy studies of diamond films usually show a sharp peak at 1332 cm^{-1}, typical for sp^3 bonded carbon (i.e., diamond), superimposed on a broad peak at about 1500 cm^{-1} which is due to graphitic sp^2 bonding. Figure 2.9 shows a typical Raman spectrum of a CVD diamond film and its optical micrograph, displaying the film morphology [2.41]. The accepted figure of merit for the evaluation of the quality of a diamond film is the ratio of the integrated intensity of the sp^3 related peak at 1332 cm^{-1} to that of the sp^2 related background. The spectrum in Fig. 2.9 was deliberately chosen to show typical results for a film of fair quality; however, excellent films with Raman spectra nearly indistinguishable from those of single crystalline diamond have been grown by many researchers. Other physical properties of synthetic diamond films, to the extent they have been studied so far, remarkably resemble those of crystalline diamond, despite their polycrystallinity and their diversity of defects (such as grain boundaries). The thermal conductivity κ of diamond films is an exception to the above, being so strongly affected by defects and grain boundaries, that $\kappa(T)$ shows a positive temperature coefficient in the range 300-700K, in contrast to the negative slope typical of single crystal diamond [2.42].

The response of diamond films to ion-implantation has not yet been extensively studied. However, from the little work done so far [2.41], it seems as if diamond films are affected by ion-implantation-induced damage in a way quite similar to that of bulk diamond crystals, despite the obvious expectation that the high concentrations of grain boundaries and defects in the films might enhance graphitization (Sect. 9.2 below).

As for the epitaxial growth of diamond films by the use of CVD methods, only successful homoepitaxy (diamond on diamond) has been reported by *Geis* [2.43]. However, a very promising ion-implantation related technique has very recently been published by *Narayan* and coworkers [2.44]. In that work, carbon ions, implanted to very high doses into single crystal Cu, have been segregated on the Cu surface in the form of a single crystalline diamond film following pulsed laser irradiation.

2.1.8 Diamond-Like Carbon Films

In addition to diamond and diamond films, there are so called "diamond-like" carbon (DLC) films which contain a mixture of both sp^3 and sp^2 bonding and have large concentrations of hydrogen impurities. DLC films are technologically important as hard, chemically inert, insulating coatings which are transparent in the infrared, and are biologically compatible with the human body [2.45, 46]. These materials, which should actually be called amorphous-hydrogenated carbon (a–C:H), are essentially a carbon based material that bridges the gap between the sp^3 bonded diamond, the sp^2 bonded graphite, and hydrogen-containing organic and polymeric materials. It has been suggested [2.47, 48] that the structure of a–C:H consists of sp^2 carbon clusters, typically planar aromatic ring clusters, which are interconnected by randomly oriented tetrahedral sp^3 bonds to hydrogen. The hydrogen atoms in DLC films may be bonded on either tetrahedral sites, where they are required to reduce bond angle disorder, or on the edges of the ring structures where they are needed for bond terminations. One major difference between DLC films and diamond (or graphite) is therefore the important role that hydrogen plays in stabilizing the DLC structure. As is shown in Sect. 9.1.3, ion implantation in a–C:H films leads to hydrogen loss, and hence to drastic changes in material properties. There is, however, little analogy between the DLC material and amorphous-hydrogenated silicon (a–Si:H) insofar as Si does not have graphitic chemistry, nor does an organic chemistry exist for Si.

The DLC films are grown by a variety of deposition techniques in which a plasma is ignited in a hydrocarbon gas mixture and the ions and radicals in it are directed towards the substrate material. Unless a very H-rich atmosphere and a heated substrate are used (in which case diamond films can be grown), the resulting film is of an amorphous structure, containing up to 40% hydrogen. These DLC films are of great commercial interest because many of their physical and chemical properties are similar to those of diamond, including hardness, chemical-inertness, electrical resistance, and some transparency in the visible and in the IR. The properties and growth methods of a–C:H have been reviewed in a number of papers and conference proceedings [2.49]. The structure and properties of the a–C:H films prior to ion implantation are

Table 2.2. Properties of a–C:H compared with those of diamond and graphite

Property	Graphite	Diamond	a–C:H
micro-structure	layered crystalline	cubic crystalline	amorphous
bonding	sp^2 Van der Waals	sp^3	sp^3/sp^2
density [g/cm^3]	2.26	3.52	1.5 –2.0
hardness [10^3 HK]	soft	10	1 – 5
electrical resistivity [Ω cm]	10^{-4}	$10^9 - 10^{12}$	10^6 10^{12}
optical band-gap [eV]	0	5.47	0.8 – 2.0

summarized in Table 2.2. The response of these films to ion implantation is discussed in Chap. 9.

An interesting diamond-like material, which is amorphous and yet contains no (or very little) hydrogen has recently been produced [2.50] by a high energy carbon ion-beam-deposition method. The carbon bonding in this material seems to be sp^3, and these films do indeed exhibit many diamond-like properties. However, the films do not show the characteristic 1332 cm^{-1} line in their Raman spectra. These films, which bridge the gap between diamond and DLC films, have so far not been used for any ion implantation studies.

2.2 Properties of Graphite

The highly anisotropic structure of graphite gives rise to lattice, transport, optical and thermal properties that are likewise anisotropic. These are briefly reviewed below for graphite prior to ion implantation. The ion implantation process will, of course, modify most of these properties.

2.2.1 Lattice Properties

Because of the strong in-plane and weak inter-planar bonding in graphite, the force constants, phonon dispersion relations, and velocity of sound are all highly anisotropic (Table 2.1). The phonon dispersion relations, phonon density of states, infrared-active and Raman-active modes of crystalline graphite have been extensively reviewed [2.51]. Of particular importance to the characterization of ion-implanted graphite is the zone center Raman-active E_{2g_2} mode with a frequency of 1582 cm^{-1} (Fig. 2.10a). In the case of 2D turbostratic graphite, this Raman-active mode at 1582 cm^{-1} persists. Also of importance to ion-implanted graphite are the maximum frequency in the phonon dispersion relations at 1630 cm^{-1} and the maximum in the phonon density of states near 1360 cm^{-1} (Fig. 2.10b) [2.51], both of which contribute to the Raman spectra

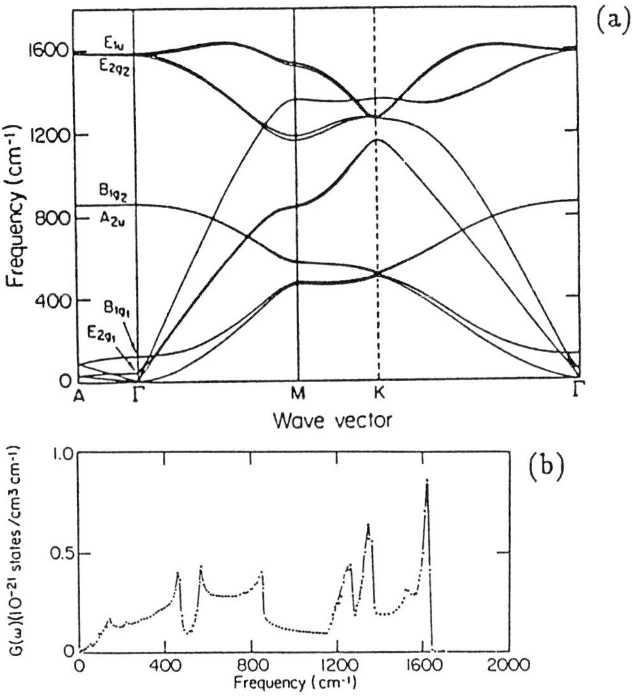

Fig. 2.10. (**a**) Phonon dispersion curves for graphite along certain high symmetry axes, (**b**) Phonon density of states corresponding to the phonon dispersion curves for graphite. The phonon frequencies are expressed in cm^{-1} [2.52]

of disordered graphite. Knowledge of the phonon dispersion relations and the phonon density of states, facilitates the use of Raman scattering for the characterization of the implantation-induced disorder in graphite-based materials (see Sect. 5.1).

2.2.2 Electronic and Transport Properties

The electronic and transport properties of graphite have been extensively reviewed [2.12, 53-55]. They are dominated by the strong in-plane sp^2 or σ-bonding and by the large anisotropy of the graphite crystal structure. Corresponding to the strong sp^2 in-plane bonding, three σ bonds are formed far below (~10 eV) the Fermi level E_F, and three antibonding σ bonds far above E_F. Thus the transport properties of graphite are largely determined by the two π-bands (one bonding and one antibonding) which lie close to E_F. By symmetry, the two π-bands formed by the p_z orbitals are degenerate at the 2D hexagonal Brillouin zone corners, and the Fermi level E_F passes through this degenerate point, giving rise to a zero gap semiconductor. This zero gap semiconductor model is appropriate for the description of turbostratic graphite (see

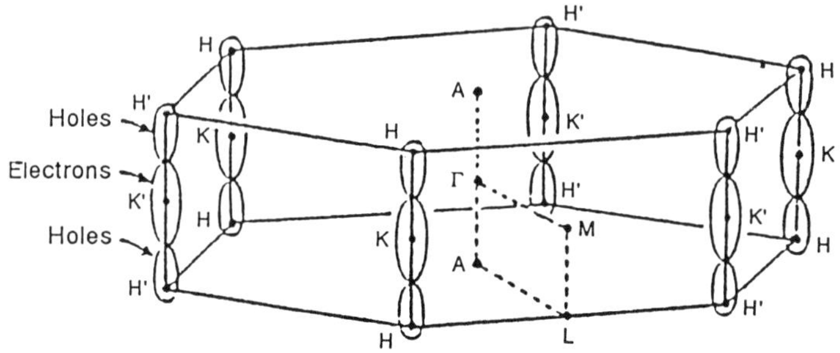

Fig. 2.11. Graphite Brillouin zone showing several high symmetry points and a schematic version of the graphite electron and hole Fermi surfaces located along the HK axes

Sect. 2.1.1) which occurs in weakly disordered graphite, as could be formed as a result of ion implantation. In contrast, perfectly ordered crystalline graphite has a weak interlayer interaction giving rise to semimetallic behavior with a band overlap of about 40 meV. The band structure near the Fermi level is described by the Slonczewski-Weiss-McClure energy band model of the four π-bands in the 3D Brillouin zone in terms of 7 band parameters. These parameters are identified with overlap integrals and transfer integrals arising from the tight binding Fourier expansion along k_z and a $\mathbf{k} \cdot \mathbf{p}$ treatment of the energy bands in the k_x, k_y directions away from the Brillouin zone edges [2.54]. Figure 2.11 shows the 3D hexagonal Brillouin zone and the location of the various high symmetry points. The band structure model near E_F focuses on the electronic dispersion relations in the vicinity of the HKH and $H'K'H'$ edges of the Brillouin zone.

The Fermi surface for semimetallic 3D graphite consists of both hole and electron carrier pockets along the Brillouin zone edges (Fig. 2.11). The small carrier pockets give rise to a low concentration of holes (p) and electrons (n), with $n = p = 3 \times 10^{18}$ cm^{-3} in the low temperature limit of an ideal graphite crystal [2.54]. The highly anisotropic Fermi surface gives rise to small effective masses for both electron and hole motion in the basal plane, but to very large effective masses for motion along the c-axis. The anisotropic Fermi surface is responsible for the high mobility [$\sim 1.3 \times 10^4$ cm^2/(V s)] for in-plane transport and for the low mobility for transport along the c-axis, with anisotropy ratios of $\sigma_a/\sigma_c \sim 10^5$ characteristic of low temperature transport in graphite [2.55]. Since the Fermi level is only 24 meV above the K-point-band extremum for electrons, there is significant thermal excitation of carriers at room temperature. Thus the temperature dependence of the in-plane electrical conductivity ($\sigma_a(T)$) is anomalous in comparison to conventional conductors and is quite different from that along the c-axis ($\sigma_c(T)$) [2.55].

The introduction of defects tends to reduce the effective band overlap (thereby reducing the carrier density) and also to reduce the carrier mobility, leading to a significant increase in resistivity from its value ($\rho_a = 41$ $\mu\Omega$ cm) in crystalline graphite [2.12]. As the defect density increases, localized states

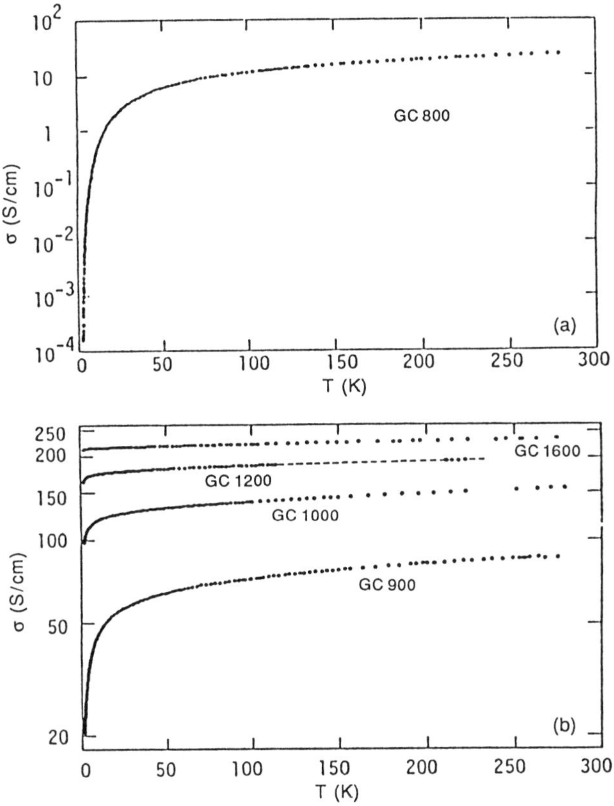

Fig. 2.12. Temperature dependence of the electrical conductivity for glassy carbon samples heat treated to various temperatures (indicated by numerals following GC) [2.28]

are formed at the respective band edges for electrons and holes, and the conductivity drops. This regime is characterized by 2D weak localization [2.56]. Eventually, when a high degree of disorder is reached, the Fermi level drops below the mobility edge (for holes), giving rise to variable range hopping conduction across an effective mobility gap. Glassy carbon is an example of such a disordered material where the conductivity has dropped by one or two orders of magnitude relative to that of crystalline graphite (4.5×10^{-3} to 4×10^{-4} Ω cm for commercial GC heat treated to $T_{HT} \sim 2500°C$) [2.23, 57]. The temperature dependence of the conductivity of glassy carbon, which is due to hopping, exhibits an $\exp[-(T_0/T)^{1/4}]$ law, characteristic of 3D variable range hopping for $T_{HT} < 1000°C$. With increasing T_{HT} $\sigma(T)$ increases in magnitude and becomes independent of temperature (see Fig. 2.12). Since the penetration depth of the implanted species is often shallow compared to the sample size, and implantation tends to reduce the conductivity as compared to that of unimplanted regions, transport properties generally do not provide a sensitive characterization tool for either the implantation-induced defects or the specific implantation-induced modification in bulk graphite-based samples.

2.2.3 Optical Properties

Graphite is a relatively good conductor; hence its optical properties are usually investigated by optical reflectivity measurements. Because of the low-carrier density of graphite, the free carriers do not give rise to a clear plasma response in the reflectivity spectrum, and thus the contributions to the dielectric function $\varepsilon(\omega) = \varepsilon_1(\omega) + i\varepsilon_2(\omega)$ from the free carriers, interband transitions and lattice vibrations all occur over the same frequency range. The frequency dependence for each of these three contributions to $\varepsilon(\omega)$ must therefore be extracted from the measured reflectivity spectra through a lineshape analysis and detailed curve-fitting procedure [2.58, 59].

Because of the high anisotropy of the electrical conductivity of graphite, the optical skin depth for the polarization **E** parallel to the *c*-axis is an order of magnitude greater than for the polarization **E** perpendicular to the *c*-axis. The resulting anisotropy of the free carrier contribution to $\varepsilon(\omega)$ together with major differences in the interband transitions for the two polarizations gives rise to a highly anisotropic optical reflectivity.

In general, optical techniques are relatively attractive for monitoring the effect of ion implantation in conducting solids because of the similar magnitudes of the optical skin depth (e.g., for wavelengths in the visible range) and the penetration depth for the implantation, both being of the order of a thousand Å. Because of the absence of any striking features in the optical spectrum of graphite at visible and infrared frequencies, the effect of implantation-induced surface damage on the reflectivity tends to outweigh the effect of reflectivity changes due to $\varepsilon(\omega)$. For this reason, the optical reflectivity has not provided a generally useful characterization tool for ion-implanted graphite. Magnetoreflection spectra, on the other hand, show well defined resonant structure that can be used to monitor the effect of ion implantation in the near surface region of graphite samples, and such applications are discussed further in Sect. 6.4.

2.2.4 Thermal Properties

The heat capacity, thermal conductivity and thermal expansion coefficients of graphite are all unusual [2.3], and yet none of these properties have played an important role in the characterization of ion-implanted graphite, the reason being the relative insensitivity of these bulk properties to the near-surface modifications introduced by ion implantation.

The unusual thermal properties of graphite (Table 2.1) have been extensively reviewed [2.3]. Ion implantation has two major effects on the thermal conductivity of graphite. The very high in-plane thermal conductivity κ_a is reduced by defect scattering, and the high degree of anisotropy ($\kappa_a/\kappa_c \approx 300$) is simultaneously reduced by the introduction of defects.

2.2.5 Mechanical Properties

Due to its strong in-plane bonding, graphite is a very stiff material with an extremely high Young's modulus (~800 GPa) and a very high in-plane tensile strength (~20 GPa). (These values for Young's modulus and tensile strength were measured on carbon whiskers [2.60].) This high tensile strength is exploited in most of the applications of carbon fibers when used in woven ropes or in composites as construction materials. Because of the weak interplanar forces, graphite has a very low shear strength ($\sim 4.8 \times 10^5$ Pa) [2.3], which, however, can be greatly enhanced through the introduction of lattice defects (e.g., through irradiation), as is further discussed in Sect. 6.5.

2.3 Properties of Diamond

The uniqueness of diamond, in contrast to all other materials having similar structures, is its instability, hence its tendency to transform into graphite when sufficient energy is supplied to the crystal. This property plays a dominant role in the response of diamond and diamond-like materials to ion implantation. A brief summary of the basic properties of diamond prior to implantation is presented below. (For more details see review papers in [2.61]).

2.3.1 Lattice Properties

The high symmetry, covalent bonding and the small interatomic distance of diamond are responsible for the unusual physical and chemical properties of this form of carbon. Because of the particularly simple crystal structure of diamond with two atoms per unit cell, its Raman spectrum is also very simple, being dominated by a single sharp zone-center, Raman-allowed line at 1332 cm^{-1}. It should be mentioned that this sharp Raman line is the main signature used to verify that in thin film carbon depositions indeed diamond, and not other carbonaceous species, has been grown. Since the Raman cross section for sp^2 bonded carbon is a factor of ~50 higher than that of sp^3 diamond, Raman spectroscopy provides a very sensitive tool for the determination of graphitic inclusions in diamond. Because of its structure and high symmetry, perfect diamond exhibits no infrared-active lattice modes.

2.3.2 Electronic and Transport Properties

Ideal diamond is a semiconductor with an indirect wide band gap of 5.47 eV. Band-structure calculations [2.62, 63] predict the conduction band minimum to fall 3/4 of the way from Γ to X in accord with experiment (Fig. 2.13). This bandgap is so wide that diamond is often considered to behave effectively as an insulator. However, as indicated above, only type IIa diamonds are pure enough

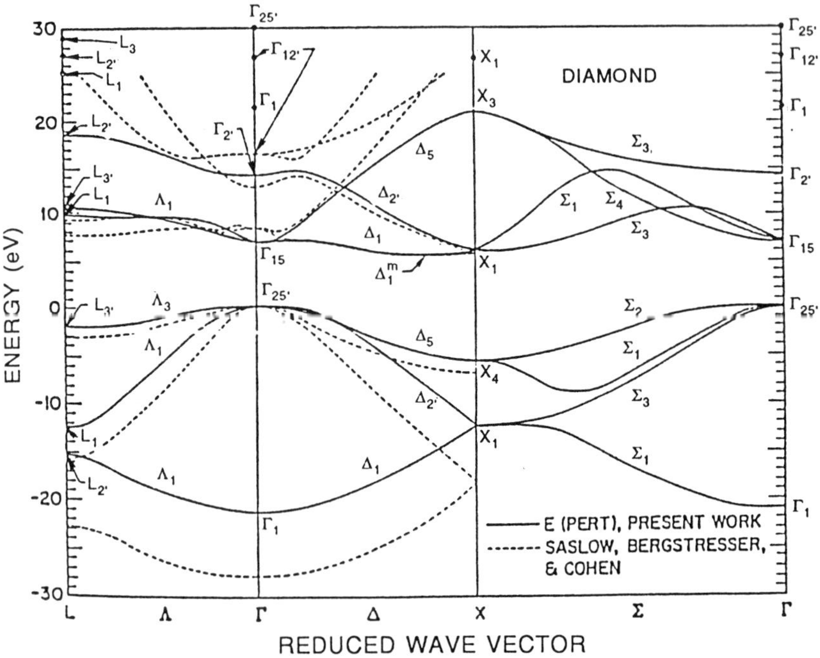

Fig. 2.13. Band-structure calculations for diamond [2.62, 63]

to exhibit the semiconducting properties of the undoped crystal. For these, resistivities exceeding 10^{16} Ω cm have been measured [2.61]. Other diamonds exhibit resistivities many orders of magnitude lower than that for type IIa, usually due to the presence of dopants. Boron with the acceptor level $\sim$0.37 eV above the valence band, is responsible for p-type conductivities in type IIb diamonds. Nitrogen, lithium and antimony have been reported to lead to n-type conductivities, however, with rather poor mobilities [2.61]. Of particular importance to the present work on ion implantation in diamond is the finding that radiation damage in diamond creates defect states which act as donors. Hence n-type conductivities can be induced in diamond by just damaging the diamond by ion-implantation. Nevertheless, the tendency of the sp^3 bonds of diamond to change to sp^2 by ion implantation and hence to form graphitic bonds may overshadow the electrical effects due to doping. Large decreases in the resistivity of ion irradiated diamond are thus usually due to the formation of highly conductive graphitic islands. The hopping conductivity between these islands is often responsible for the low resistivities which are measured in ion implanted diamond and are sometimes erroneously attributed to chemical doping of diamond.

2.3.3 Optical Properties

One of the major attractions of diamond as a gem stone, ever since diamond was first discovered and polished, is its well known glitter. This is caused by the high index of refraction of diamond ($n = 2.42$ in the visible) and by its low optical absorption. Actually diamond is not only transparent in the visible but also over most of the infrared and ultraviolet regions of the spectrum (Fig. 2.14). The major absorption in the IR (at 7-8 μm) and in the UV (at $\sim$0.25 μm) in natural diamond is due to the presence of nitrogen impurities as shown in Fig. 2.14 [2.64].

Most studies on the optical properties of diamond have used absorption and luminescence techniques. These techniques were developed among others by the need to devise a non-destructive simple and quick way to evaluate the purity of natural diamond. Nitrogen, and nitrogen complexes have been found to exhibit a distinct optical fingerprint, and were thus most extensively studied. The effects that irradiation have on the optical properties of diamond have been studied, with the aim of modifying the color of the stone at will; hence, deeply penetrating radiation, such as UV photons, electrons or neutrons have been used. It was found that MeV electron irradiation to high doses ($> 10^{18}$ cm^{-2}) gives rise to a blue-green appearance of the diamond, probably due to optically active complexes formed between irradiation-induced point defects and nitrogen impurities. Neutron irradiation has been shown to cause similar changes in the appearance of irradiated diamonds. Several absorption lines, ranging in energy from $\sim$1.6 to 3.0 V (and denoted by "GR"), have been observed in many irradiated diamonds, and are believed to be related to various excitation states associated with energy levels due to neutral vacancies in the forbidden gap of diamond. Many of the optically-detected defects can be "bleached", i.e., can be either thermally or optically annealed.

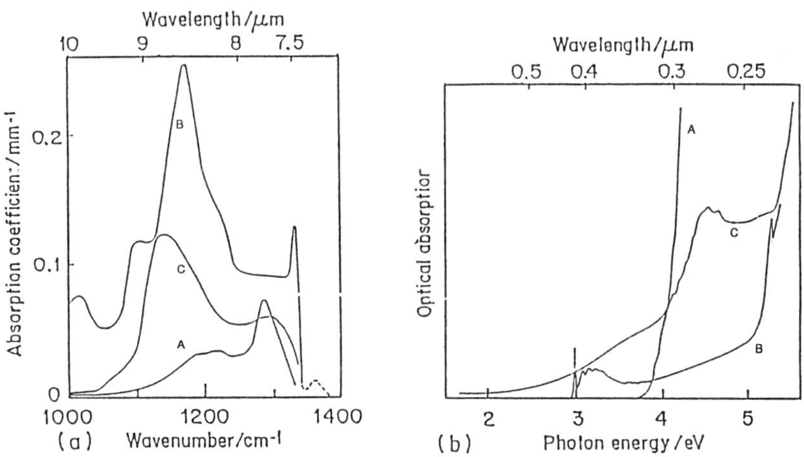

Fig. 2.14. (**a**) Infrared absorption spectra and (**b**) UV absorption spectra of type I diamonds containing nitrogen as different agglomeration forms (curves A and B) and as substitutional atoms (curve C) [2.64]

The effects that ion-irradiation have on the optical properties of diamond are more complex, and have been studied much less. The complexity in this case stems from the fact that bombardment by heavy projectiles leads to a very high density of damage centers created during the slowing down of the implanted ion (as will be discussed below, see Sect. 3.3). The point defects caused by the damage cascade around the projectile track tend to agglomerate to form defect complexes, the optical properties of which have not been extensively studied. Low dose ion irradiation causes the diamond to exhibit a yellowish appearance which gets darker with increasing dose until eventually the irradiated region looks dark brown or even black. This is probably due to the formation of graphitic regions in the heavily damaged layers, as will be discussed later (Sect. 8.1).

2.3.4 Thermal Properties

The thermal conductivity (κ) of diamonds is (along with in-plane κ_a in graphite) the highest of all materials [2.65]. It peaks at ~80 K, where the conductivity of some type IIa diamonds was found to be as high as 1.5×10^5 W $m^{-1}K^{-1}$. Even though κ drops off at higher temperatures, it still exceeds that of copper by about a factor of 6 at room temperature (Fig. 2.15). This makes diamond a most attractive material for both microelectronics applications and for cutting tools, since in both these applications fast heat dissipation is most important.

The thermal conductivity is much reduced by crystal imperfections and by defects, which act as phonon scatterers, so that κ strongly depends on diamond

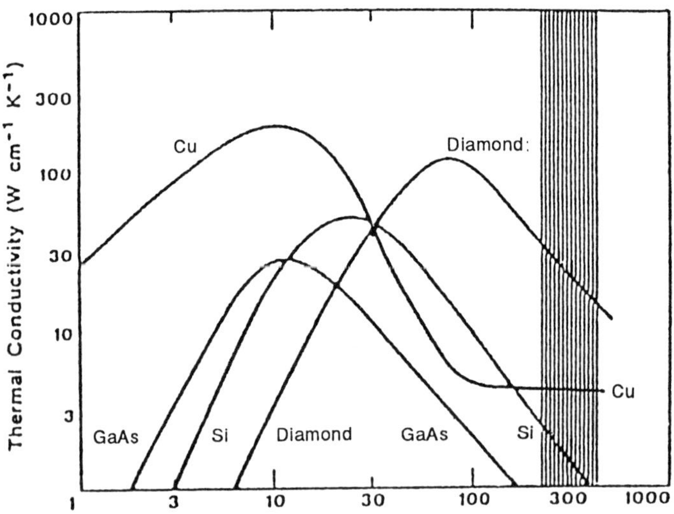

Fig. 2.15. Thermal conductivities of several solids (including a IIa diamond) as a function of temperature. The group of vertical lines around 300 K represents the temperature range from −25° to 125°C [2.65]

type (through the presence of impurities) and on radiation damage. Typical type I diamonds may exhibit thermal conductivities three times lower than those of good type IIa stones, but nevertheless still higher than that of copper and other excellent thermal conductors. It should be mentioned, that very recently, isotopically enriched diamonds (containing ~99.9% of ^{12}C) have been artificially synthesized [2.66]. These most defect-free crystals have exhibited thermal conductivities over a factor of 2 superior to any previously measured κ. The relevance of thermal conductivity to the evaluation of ion implantation effects has not been exploited at all, the main reason being the shallowness of the implantation-affected regions (~0.1-0.5 μm), which makes measurement of the thermal properties of a thin layer on top of a bulk, highly conductive substrate, very difficult.

The specific heat of diamond, a crystal with a very "stiff" bond, is extremely high (~6.2 Jg^{-1}), corresponding to Debye temperatures of ~2000 K at room temperature [2.61]. The thermal expansion coefficient of diamond is $\sim 1 \times 10^{-6}$/K at 300 K and increases up to $\sim 5 \times 10^{-6}$/K at 1200 K. In contrast to graphite, which exhibits high anisotropy in all thermal (and other) properties, diamond, having a highly symmetric structure, is isotropic in all properties which can be represented by a second rank tensor.

2.3.5 Mechanical Properties

Diamond is extremely hard in all respects. The indentation hardness, as well as scratch hardness are the highest of all known materials, somewhat depending on crystal orientation. Interestingly, even though ion-implantation modifies most of the physical properties of diamond and makes them more "graphite-like", implanted diamond retains many of the superb mechanical properties of the unimplanted material which can, under certain implantation conditions, be further improved (Sect. 8.9).

2.3.6 Chemical Properties

Diamond is completely resistant to all chemical etchants at room temperature. In that respect diamond differs from graphite which can be etched by some very active boiling mixtures of acids (for example, 1:1:1 of HNO_3; H_2SO_4; $HClO_4$). This difference between graphite and diamond surfaces can be exploited to remove graphite from diamond. As for reactive ion etching, graphite etches more easily by atomic oxygen or hydrogen than diamond, a fact which is probably responsible for the preferential growth of diamond films by the various CVD methods, all of which require the presence of large amounts of H atoms or some oxygen in the plasma to promote the sp^3 bonding.

When diamond is subjected to a sufficiently high local energy density, as is the case for heavy ion impact, some of its sp^3 bonds may transform into sp^2 graphite-like bonds, thus altering many of its physical and chemical properties. These changes will be discussed in detail in Chap. 8.

3. Ion Implantation

Ion implantation is a technique which enables one to introduce almost any impurity (ion species) into the near surface region of any solid, including diamond, graphite or any other carbon-based material. In this section the energy loss processes of the incoming energetic ions are discussed in general, ending with some reference to carbon-based materials (Sect. 3.1). The various parameters of the implantation process are then introduced in general, and specific values are given for carbon materials (Sect. 3.2). The implantation-induced damage is described in Sect. 3.3. The chapter concludes with a discussion of the energy loss process (Sect. 3.4) for ion implantation into carbon materials. A more complete treatment of the general topics presented in this chapter can be found in References [3.1-3].

3.1 Energy-Loss Mechanisms

An energetic ion penetrating into any solid loses its energy through scattering events involving the Coulomb interaction of the ion with the atoms and electrons in the target. This energy loss determines the final penetration of the projectile into the solid and the amount of disorder that is created in the lattice. The ion-solid interactions which describe the physics of the energy loss of ions slowing down in a solid have recently been reviewed by *Eckstein* [3.4].

Each collision of the projectile with constituents of the target is a complicated many-body event described by a complicated Hamiltonian, which is usually approximated by assuming that the interaction between the ion and the scatterer can be separated into two major components, namely an ion (projectile)-nucleus (target) interaction and an ion (projectile)-electron (target) interaction.

The slowing down process of an ion in a solid can thus be roughly divided into two dominant mechanisms for energy loss by the incident ions, namely electronic energy loss, involving the interaction between the incident ion and electrons of the host material, and nuclear energy loss, involving the interaction between the incident ions and the atoms (considered as shielded point charges) of the host material.

The general way to treat the slowing down of an ion in matter is through the "stopping power" (dE/dx), defined as the energy dE lost by an ion traversing a distance dx. The stopping power is given by

$$\frac{dE}{dx} = N \int T d\sigma \tag{3.1}$$

where $d\sigma$ denotes the collision cross section, T is the energy lost by the ion in the course of a collision event, and N is the density of scattering centers of the host material. The stopping cross section $\mathcal{E}$ is thus given by

$$\mathcal{E} = \frac{1}{N}\frac{dE}{dx} = \int T d\sigma. \tag{3.2}$$

The total stopping power is, as mentioned above, due to both electronic and nuclear processes and can thus be written as

$$\frac{dE}{dx} = \left(\frac{dE}{dx}\right)_e + \left(\frac{dE}{dx}\right)_n = N(\mathcal{E}_e + \mathcal{E}_n) \tag{3.3}$$

where $\mathcal{E}_e$ and $\mathcal{E}_n$ are the electronic and nuclear stopping cross sections, respectively.

A detailed description of the electronic energy loss has been given by *Lindhard* et al.[3.3], in the commonly called LSS theory. In this approach the electrons in the solid are viewed as a free electron gas which is affected by the traversing positive charge of the projectile. The inhomogeneity in the electron plasma introduced by the moving ion induces a stopping force on the ion, and hence a loss in the projectile energy. This approach predicts for ion velocities of relevance to ion implantation that the electronic stopping is proportional to the ion velocity, i.e., $(dE/dx)_e$ is proportional to $\sqrt{E}$. Lindhard's model does not allow any deviation in the projectile trajectory for ion-electron interactions.

In contrast, the ion-nucleus interaction is given by the simple Coulomb interaction between two positive (screened) charges which results in both energy loss and significant deviation in the projectile trajectory. The cross section for this process is given by the simple Rutherford scattering cross section [(4.1) below]. In the ion-nucleus interaction, not only do the colliding ions change directions, but the atoms of the host material may also be significantly dislodged from their original positions, giving rise to lattice defects. The deviations in the projectile trajectory results in both a lateral spread and a depth distribution of the implanted species, while the displacements of the host atoms gives rise to lattice damage.

Simple kinematic calculations show that the energies of the projectile before (E_0) and after (E_1) scattering are related by [3.1, 2]

$$E_1 = kE_0 \tag{3.4}$$

where the kinematic factor k is given by

$$k = \left(\frac{M_1 \cos\theta \pm (M_2^2 - M_1^2 \sin^2\theta)^{1/2}}{M_1 + M_2}\right)^2 \tag{3.5}$$

and the scattering angle is given by

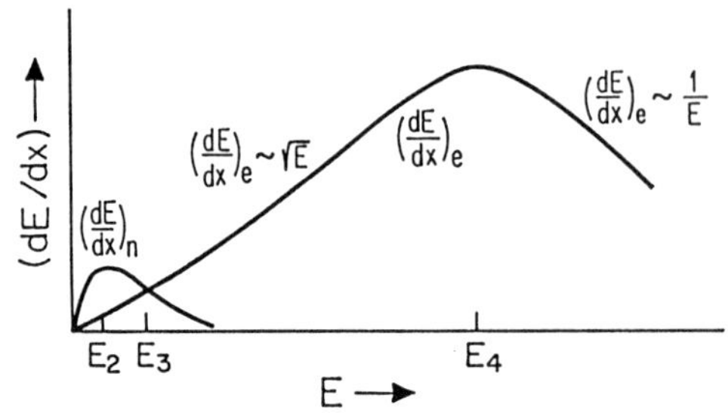

Fig. 3.1. Dependence of the nuclear $(dE/dx)_n$ and electronic $(dE/dx)_e$ energy loss (stopping power) on the energy of the projectile

$$\cos\theta = \frac{1-(1+M_2/M_1)(T/2E_0)}{\sqrt{1-T/E_0}}. \tag{3.6}$$

Here M_1 and M_2 are the masses of the projectile and target atoms, respectively, θ is the scattering angle of the projectile in the laboratory frame of reference, while T is the recoil energy transferred from the projectile to the target, which is determined from the conservation of energy, i.e.,

$$T = E_0 - E_1. \tag{3.7}$$

Relations (3.4-7) are completely general no matter how complex the force between the two particles, as long as the force acts along the line joining the particles and the collision can be taken to be elastic. In the ion-electron interaction, $\theta \sim 0°$ and the maximum recoil energy is $T_{\max} = 4mE_0/M_1$ in which m and M_1 are the masses of the electron and the ion, respectively. This interaction induces small losses in the energy of the incoming ion as the electrons in the host atoms are excited to higher bound states or to ionization states. These ion-electron interactions do not produce significant deviations in the projectile trajectory.

From the energy loss, the ion range can be calculated according to

$$R = \int \frac{dE}{dE/dx} \tag{3.8}$$

where the integration limits are from the initial ion energy to zero. From knowledge of the energy transferred to the electrons and to the lattice (including both phonon generation and permanent displacements of the host ions), the energy of the incoming ions can be calculated as a function of distance along their trajectory $E(x)$.

At low energies of the projectile ion, nuclear stopping is dominant due to the $1/E^2$ dependence of the Rutherford cross section [(4.1) below], while electron stopping dominates at high energies, as shown in the characteristic stopping power curves of Fig. 3.1. Three important energy parameters are indicated in

Fig. 3.1: E_2 is the energy where the nuclear stopping power is a maximum, E_4 where the electronic stopping power is a maximum, and E_3 where the electronic and nuclear stopping powers are equal. As the atomic number of the ion increases for a fixed target (e.g., carbon), the scale of E_2, E_3 and E_4 increases. Also indicated on the diagram is the functional form of the energy dependence of the stopping power in several of the regimes of interest. Typical values of the parameters E_2, E_3, and E_4 for the implantation of carbon into diamond are: E_2 [maximum of $(dE/dx)_n$]<5 keV; E_3 [where $(dE/dx)_n = (dE/dx)_e$]$\sim$ 15 keV; E_4 [maximum of $(dE/dx)_e$]$\approx$ 2 MeV. It is important to remember that only in the energy regime in which nuclear stopping dominates are the host atoms severely dislodged. Electronic stopping will usually not create extensive damage to the host crystal, except in those cases for which electronic excitation and bond breaking are possible, so that some structural rearrangement may take place.

The slowing down processes of an ion moving in an amorphous solid is a statistical process. Thus the locations at which the implants finally come to rest are also of a statistical nature, and should be expressed using statistical variables. The ion range R is related to the mean track length of the ion before coming to rest, while the projected range R_p gives the mean penetration depth of the ion relative to the surface [3.5]. Considering the projected range distribution, the Gaussian approximation gives the implant density $n(x)$ as

$$n(x) = \frac{\phi}{\sigma_p\sqrt{2\pi}} \exp\left[-\frac{(x-R_p)^2}{2\sigma_p^2}\right] \tag{3.9}$$

where x is measured along the direction of the beam, ϕ is the fluence, or the ion dose, while σ_p is the standard deviation in the projected range R_p, (also referred to as the ion range straggling ΔR_p). This Gaussian approximation is useful for gaining a simple physical picture of the implant profile. It should be noted that the profile given by (3.9) ignores diffusion which may take place during the implantation and higher order effects, as well as channeling, to be discussed in Sect. 4.4.

More realistic information about the implant distribution in the target and the related damage inflicted is obtained from computer simulations discussed in Sect. 3.4. Detailed calculations of the ion energy loss and of the final ion distribution have been carried out for the implantation of various ion species into carbon host materials, without taking into account any specific order or structure of the target other than its density; these are discussed further in Sect. 3.4. The slowing down of ions along crystallographic directions in single crystals under the so-called channeling conditions is treated in Sect. 4.4.

3.2 Parameters of Implantation

The discussion in Sect. 3.1 of the energy loss of the energetic implanted ions as they come to rest in a given host material has identified a number of the important implantation parameters, including the ion species (of mass M_1), its energy E_0, the mean ion range R, the mean projected range R_p, the projected range straggling ΔR_p, the fluence (or dose) ϕ, and the flux (or beam current) i_b.

3.2.1 Energy of Implantation

The interaction of ion beams with solids depends on the energy, the mass and the charge of the incident ion, and on the properties of the stopping medium. In the very low energy range, a directed low energy ion ($\sim$ 10–100 eV) comes to rest at or within a few atomic layers of the surface of a solid, giving rise to an ion beam deposition process, possibly growing into a registered epitaxial layer upon annealing. At a somewhat higher energy (e.g., 1 keV), a heavy ion beam is often used for sputtering, that is, removing a very thin layer of material from the near surface region. For this application, rare gas ions such as Ar or chemically reactive ions such as oxygen or cesium are normally used, and a large fraction of the incident energy is transferred to the atoms of the solid. This results in the ejection of surface atoms into the vacuum, leaving the surface in a disordered (roughened) state. Sputtering has been used to remove damaged edge surfaces normal to the layer planes of graphite. Sputtering is also used for depth profiling of the implant distribution, since it allows very gradual "digging" into the material while probing the ejected species using the secondary ion mass spectrometry (SIMS) or Auger electron spectroscopy techniques [3.2] (Sects. 4.6 and 5.3).

Incident ions in the range $\sim$ 10–300 keV are used in the ion implantation process to modify the properties of carbon-based materials to a depth of $\sim 10^2 - 10^4$ Å (Table 3.1) depending on the ion mass and target density. Because of the differences in density in the various carbon-based materials, Table 3.1 can be used quantitatively only for diamond host materials. Use of high energy (above $\sim$ 300 keV) light projectiles (H or He ions) allows the study of the distribution of the chemical constituents and the lattice disorder as well as epitaxial layer formation by the application of various ion beam analysis techniques, as is discussed further in Chap. 4.

Table 3.1. Values of implantation parameters for various ion species implanted into diamond at 100 keV, based on TRIM calculations (Sect. 3.4). The parameter for the displacement energy was taken as $E_d = 55$ eV

Ion	R_p[Å]	ΔR_p[Å]	Vacancies/ion
B	1744	262	87
C	1388	223	94
P	670	163	207
As	374	89	283
Sb	286	55	310
Bi	221	30	294

3.2.2 Implantation Range

As defined by (3.8), the mean range of an implanted ion (such as carbon ions with energies between 100 and 300 keV implanted into a carbon host) is on the order of 1000–3000 Å, so that ion implantation must be considered as a near surface phenomenon. The projected mean range R_p defined by (3.9) is the actual quantity of interest in defining the peak position, as measured from the target surface, and the corresponding half-width at half maximum ΔR_p of the implant distribution. Some typical values for R_p and ΔR_p are given in Table 3.1, which also includes the mean number of vacancies per implanted ion, to be further discussed in Sect. 3.4. For implants of a given incident energy, high mass ions will come to rest close to the surface, and the lower the mass, the larger the penetration depth R_p, as follows from considerations of conservation of energy and momentum in the implantation process.

3.2.3 Implantation Fluence (Dose) and Beam Current (Dose Rate)

In addition to the intrinsic parameters which affect the implantation process discussed above, there are the extrinsic parameters, the ion fluence and the beam current. The ion fluence or ion dose denotes the total number of ions incident per unit area of the target material. In many cases in which ion-implantation is used to modify the electronic or optical properties of the affected near-surface regions (typically several thousand Å thick), implant concentrations on the order of parts per million or less are required. These correspond to implantation fluences of the order of 10^{12}-10^{14} ions/cm^2. However, for those applications in which major structural modifications of the implanted surfaces are needed, large implant concentrations (on the order of $\sim$1%) must be achieved, requiring implantation to very high fluences ($10^{16} - 10^{17}$ cm^{-2}). For such high implantation doses, the solid solubility limit of the foreign atom forced into the host matrix may often be exceeded. Furthermore, for low-energy, heavy-ion implantations, severe surface erosion due to sputtering may also occur. The physical properties of ion-implanted materials are sensitive to the ion fluence ϕ, largely because of the lattice damage associated with the implanted ions (Sect. 3.3). Annealing is often necessary to reduce the lattice damage so that the effect of the implant as a charged or neutral impurity species can be realized.

Also of significance is the rate at which the implantation occurs. Slow implantation, characterized by a slow arrival rate of the ions at the surface, may allow the system to partially recover from the radiation damage before further implantation-induced damage is inflicted to the same volume in the target. In contrast, fast implantation may cause overlap of damage cascades and hence result in a different kind of damage. If no particular care is taken, ion implantation may lead to a substantial rise in target temperature due to the high power injection into the specimen by the ion beam. The implantation rate is often expressed in terms of the beam current density i_b in units of μA/cm^2. Values in the range of a few μA/cm^2 or less are typical for slow implantation, while dose rates of 100 $\mu A/cm^2$ or more are needed to achieve high dose implantation within reasonable times.

Another variable of importance is the temperature of the substrate during implantation, which may either be selected by the experimenter or may be different from the temperature of the substrate prior to implantation because of the heating due to the power delivered by the ion beam. Increasing the implantation temperature substantially increases the diffusion rate of both implants and defects, and thus increases the probability of recombination of vacancies and interstitials, thereby annealing the lattice damage, while at the same time spreading out the implantation profile. Implantation at low temperature is therefore desirable for maintaining a sharp impurity profile at the expense of "freezing in" the damage, while implantation at elevated temperature is desirable for instantaneous annealing of the lattice damage.

3.3 Radiation Damage

The energy transferred from the projectile ion to the target atom is usually sufficient to break a chemical bond and to cause a displacement of the target (carbon) atom from its original lattice site. The condition for this process requires the energy transfer per collision to be greater than the displacement energy E_d. Because of the high incident energy of the projectile ions, each incident ion can dislodge multiple-host ions. Each one of these may have sufficient energy to inflict further damage during its slowing down process. Thus the slowing down of a single implanted ion in the solid will usually initiate a whole cascade of atom displacements; hence the terms "damage cascade" and "thermal-spike" which are used to describe this process. The number of vacancies created per incident ion may thus be on the order of hundreds or even thousands, and is an important characterization parameter of the radiation damage (Table 3.1). As the ion slows down, its energy decreases, and since $(dE/dx)_n$ increases with decreasing ion energy (due to the $1/E^2$ dependence of the Rutherford scattering cross section), the ion will interact more strongly with the lattice of the host and will dislodge more atoms, thereby creating a higher density of vacancies. The damage profile for low dose implantation gives rise to isolated regions of damage within the damage-cascade volume, but as the fluence is increased, these damaged regions overlap (Fig. 3.2) and may eventually coalesce to form a new phase. Such is the case for ion-damaged diamond which eventually is transformed into graphite.

The damage profile also depends on the mass of the projectile, with heavy mass ions of a given energy causing more local lattice damage as the ions come to rest. Typical damage patterns are shown in Fig. 3.3 on the basis of TRIM calculations (Sect. 3.4) for a light ion (carbon) and for a heavy ion (antimony) slowing down in diamond.

To compare the amount of lattice damage produced by ions of different masses and energies, experimental data are often plotted in terms the number of displacements per atom [dpa], a number which is obtained by dividing the

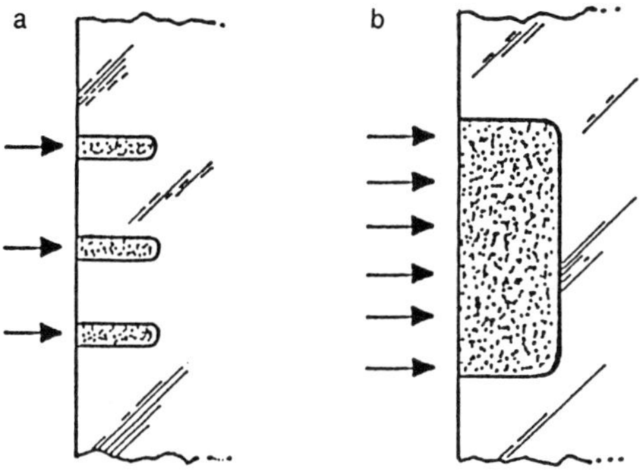

Fig. 3.2. Schematic diagram for damage cascade for (**a**) low dose, and (**b**) high dose implantation

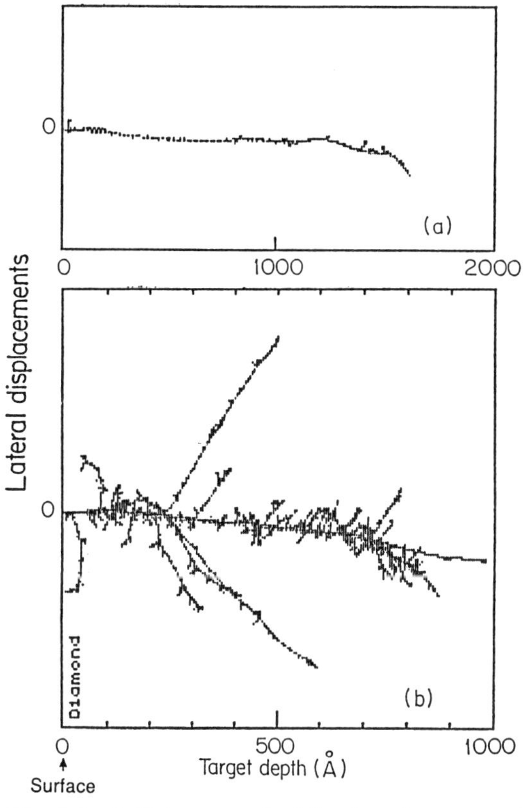

Fig. 3.3. Collision cascade following the penetration of (**a**) a single 100 keV $^{12}C^{+}$ ion or (**b**) a single 300 keV $^{121}Sb^{+}$ ion in diamond (from TRIM simulations). The darker track (starting at the origin) denotes the track of the primary ion while the lighter tracks represent the trajectories of carbon recoils (primary as well as higher generations)

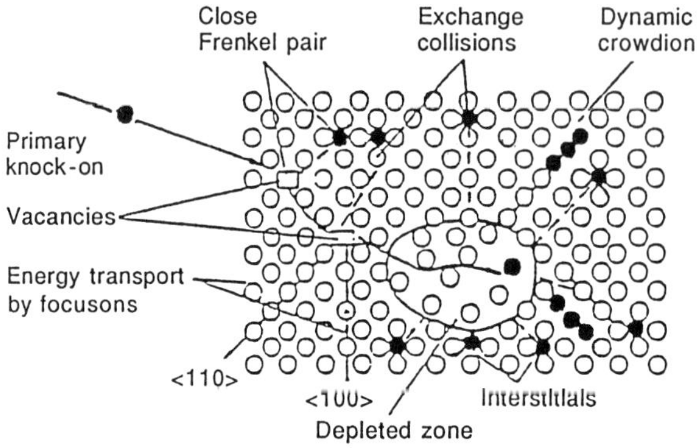

Fig. 3.4. Example of typical defects induced by ion implantation

defect density D^* by the atomic density N of the host material (displacements per atom [dpa]= D^*/N). Here D^* is obtained by multiplying the dose ϕ by the vacancy production per ion (vpi) and then dividing by V_d the volume of lattice damage, to obtain $D^* = \phi(\text{vpi})/V_d$. Explicit values for D^* are determined from simulation programs such as TRIM, described in Sect. 3.4.

A schematic diagram of the types of defects caused by ion implantation is shown in Fig. 3.4, which illustrates among others the formation of vacancies, interstitials and simple complexes such as Frenkel pairs (the pair formed by a vacancy and an interstitial). Once the density of elementary point-defects is high enough, they may agglomerate to form defect complexes, generally called extended defects. These may include dislocation loops and stacking faults. The formation of extended defects may be enhanced by elevating the temperature during implantation (usually referred to as "hot implantation"), and conversely extended defect formation may be suppressed by low temperature, low current density implantations, in which case only the "frozen-in" elementary point defects will be present.

In order to reduce radiation damage and to anneal the crystal, energy must be supplied to the system so as to enable the diffusion of interstitials and vacancies and to allow their mutual annihilation or to assist the break up of extended defects. The required energy is usually supplied in the form of heat provided by furnaces, lasers or flash lamps. The local energy provided to the target during the implantation process itself, especially when combined with heated substrates, may, under certain circumstances, assist damage removal, giving rise to a phenomenon called "ion beam-induced annealing".

3.4 Energy-Loss Simulations

Computer simulations of ion-solid interactions have recently been reviewed in a book on that subject by *Eckstein* [3.4]. Only the most commonly used simulation is briefly described below.

To obtain a quantitative description of the energy loss of the energetic ions, the simplistic approach of Sect. 3.1, where the electrons and nuclei of the host material are described as free particles, must be refined by introducing realistic interaction potentials for the electrons and the nuclei. Such calculations have been carried out since 1963, starting with the pioneering work of *Lindhard, Scharff and Schiøtt* (LSS theory) [3.3], and have been continuously refined. With the ready availability of computers, the emphasis has turned to the generation of useful computer codes to simulate the energy loss process. The most commonly used code today is the TRIM program (Transport of Ions in Matter) developed by *Ziegler* et al. [3.6, 7] in which the best known semi-empirical values for the stopping powers as well as the statistical nature of the slowing down process are incorporated. The TRIM program is a Monte-Carlo program which exactly follows the collisions that individual ions undergo while in motion in the target material. The input parameters to the program are the required experimental conditions (ion type and energy, target material, composition and density) and the intrinsic parameter E_d, which is the energy required to displace a target atom far enough from its lattice site so that it will not fall back into the vacancy that it has left behind. This parameter is usually not well known. An uncertainty in E_d is directly reflected in uncertainties in the calculated numbers of vacancies and interstitials.

The procedure followed by the program is to shoot the required ion into the target; the ion will then proceed into the target unaffected until it collides with a target atom with a randomly selected impact parameter. The TRIM program calculates the kinematics of the collision, and then follows the trajectory of the recoiling target atom which itself undergoes collisions that may put new target atoms in motion, thus creating a branch of recoiling target atoms. The program will return and follow the motion of the original implant once all recoiling target atoms have attained energies below a certain cutoff energy E_{co}. The projectile may now proceed with its new energy and direction of motion until another statistically selected collision occurs with its branch of recoils. The procedure continues until the energy of the implant itself is below E_d, and hence the implant cannot dislodge any more target atoms and it comes to rest. The program now stores the collision history of the primary ion and of all secondary events and proceeds to follow a new ion shot into the target.

Typical damage cascades created by a single light (C) and a single heavy (Sb) ion implanted at 100 keV and 300 keV, respectively into diamond are shown in Figs. 3.3a and b. Figures 3.5a and b show the implant and recoil distributions after 310 carbon or 122 Sb ions have been injected into the target at the origin. The mean ion ranges and their distributions (which are not exactly Gaussian) can be obtained from these simulations. Figures 3.6a and b show the statistical distributions of the implanted ions and of the vacancies created by the above C and Sb implantations into diamond. It is important to note the large difference in the average number of vacancies created by each carbon or antimony ion, being 110 and 870 vacancies per ion, respectively. Hence, the total damage and the density of the damage cascades of heavy ion implants are much larger than for light ion implants.

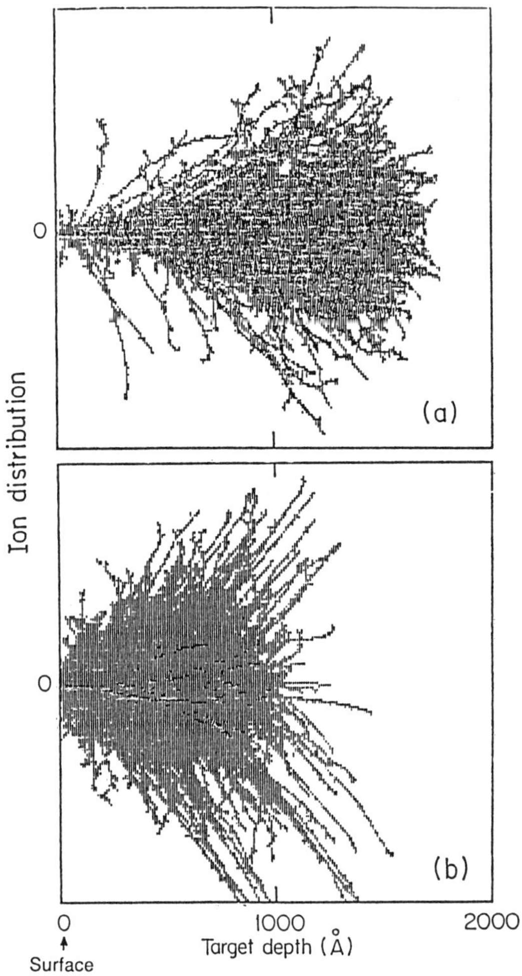

Fig. 3.5. Tracks obtained from the TRIM program for (**a**) 310 $^{12}C^+$ ions implanted at 100 keV and (**b**) 122 $^{121}Sb^+$ ions implanted at 300 keV into diamond. The dark spots denote the resting positions of the primary ions, all of which start off at the origin, and are directed normal to the surface. The lighter tracks represent the trajectories of higher generations of recoils. Compare these tracks to that of a single ion shown in Fig. 3.3

Even though the TRIM simulation gives excellent insight into the implantation process and the damage that accompanies ion implantation, the simulation program ignores several features which are of importance in the real implantation process. This program for example assumes a statistically homogeneous distribution of the host atoms, according to the average density of the target material; it is therefore not capable of properly handling anisotropic materials, like graphite, and very misleading results may be obtained in such cases. Fur-

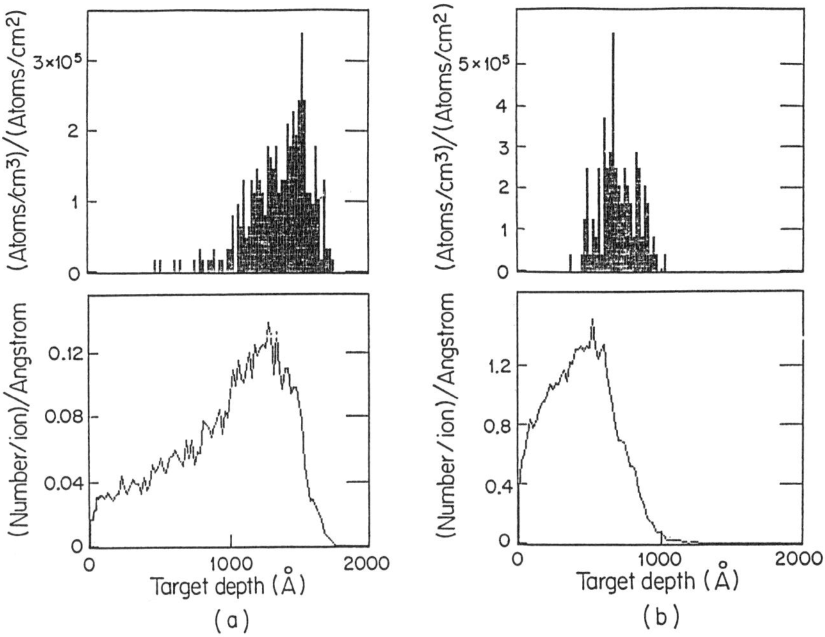

Fig. 3.6. Statistical distribution of implants and of vacancies corresponding to (**a**) 100 keV $^{12}C^{+}$ ions and (**b**) 300 keV $^{121}Sb^{+}$ ions implanted into diamond. Plots are shown for both the ion range of the implants (*top figures*) and the collision events which relate to the vacancies and total damage that is produced (*bottom figures*)

thermore, effects which are due to the periodicity of atoms in a single crystal (channeling) are neglected by the program. TRIM also does not allow for any recombinative annihilation of defects (i.e., no vacancy-interstitial recombination is included in the program), so that TRIM always overestimates the absolute magnitude of the damage. The program also does not allow for the rearrangement of atoms in the implantation-affected volume to form new phases, such as the conversion of diamond into graphite. Nevertheless, TRIM and similar computer simulation programs give excellent estimates of both the implant and the damage profiles, and hence offer very valuable information on the state of a target after ion implantation.

4. Ion Beam Analysis Techniques

A variety of experimental techniques have been used to characterize the implantation-induced modifications to the target material, as well as the associated lattice damage. In this chapter, some background is provided on the most important characterization techniques based on ion beam analysis (IBA).

The interactions that charged particles experience, while in motion in a solid, can be utilized to gain information about the structure of the material under study. However, in contrast to ion implantation, in which the major effect on the solid is its near-surface modification due to the introduction of impurities or to damage to the lattice, most ion beam probing techniques should be as "non-destructive" as possible. Hence, the ion beams used for this purpose are usually selected so as to minimize damage to the material under study. This requirement limits the beams useful for ion beam analysis (IBA) to ions for which electronic stopping dominates, i.e., light projectiles (e.g., H, He) in the MeV energy range.

The principle which is common to most ion beam analysis techniques is the detection of a signal which characterizes a short range interaction between the probing projectile and a target atom. This signal may be a back scattered probing particle, an X-ray characteristic of the target atom excited by the projectile or products typical of a nuclear reaction between the probing ion and a target nucleus.

The information which can be obtained from IBA experiments is diverse. It ranges from information on impurities in the material, on thin film deposits on the sample surface, and when applied to single crystals in the so-called channeling experiments, information can be obtained on the degree of perfection of the crystal structure and about the location in the lattice that is occupied by foreign atoms. As most of this information is essential for the assessment of implantation-induced effects on the target materials, it is not surprising that IBA techniques have been widely used also in implantation and annealing studies in the carbon-based materials discussed here.

In the following sections of this chapter the basics of the various ion beam analysis techniques are briefly reviewed. The most commonly used Rutherford backscattering spectroscopy (RBS) technique is described in Sect. 4.1; brief surveys of the nuclear reaction analysis (NRA) and of the particle-induced X-ray emission (PIXE) technique are given in Sects. 4.2 and 4.3, respectively. Channeling is discussed in Sect. 4.4 and the elastic recoil detection method in Sect. 4.5. A completely different probing technique, which also utilizes ion beams but is based on sputtering, is the secondary ion mass spectroscopy (SIMS) method. Here, in contrast to the other IBA techniques, predominantly nuclear stopping is required; hence, low energy (keV) heavy ions (e.g., ^{16}O, ^{132}Cs) are used and the atomic and molecular species sputtered from the target surface are mass analyzed. The method is briefly described in Sect. 4.6.

For more information on the various ion beam analysis techniques, the reader is referred to text books which review the field [4.1-5].

The application of RBS spectroscopy and channeling to characterize the most extensively used form of graphite for research purposes (highly oriented pyrolytic graphite) is presented in Sect. 4.7. The application of RBS spectroscopy to graphite intercalation compounds for compositional analysis is described in Sect. 4.8 and for structural alignment using channeling techniques in Sect. 4.9.

4.1 Rutherford Backscattering Spectroscopy

In Rutherford backscattering spectroscopy (RBS) [4.1], a beam of monoenergetic (usually 1-2 MeV), collimated, light mass ions (H^+, He^+) impinges on a target and the number and energy of the particles that are scattered backwards at an angle θ (usually close to 180°) are monitored (Fig. 4.1). This simple experiment can yield information about the composition of the target (host species and impurities) as a function of depth.

Compositional information is obtained for different mass atoms ($M_2^{(i)}$) located on the surface of the target from the different backscattered energies $E_1^{(i)}$ that the probing particles (with mass M_1) assume following a Rutherford scattering event through an angle θ. These energies are given by (3.4) and (3.5) discussed in Chap. 3. It follows from (3.5) that in order to obtain high mass

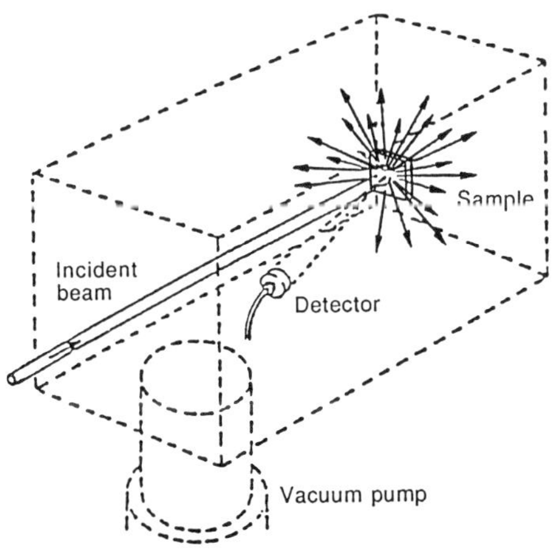

Fig. 4.1. In the Rutherford backscattering experiment, the scattering chamber, where the analysis is actually performed, contains the following essential elements: the sample, the beam collimation system, the detector, and the vacuum pump [4.1]

resolution in an RBS experiment, i.e., to be able to discriminate between atoms of different masses, the kinematic factor k, (3.5), should be as large as possible. This occurs for θ close to 180° and for large projectile masses M_1 (as long as $M_1 < M_2$). The differential cross section $d\sigma/d\Omega$ for an elastic collision by which the projectile is deflected by an angle θ is given in the laboratory frame by

$$\frac{d\sigma}{d\Omega}=\left(\frac{Z_1 Z_2 e^2}{4E}\right)^2\frac{4}{\sin^4\theta}\frac{\{[1-((M_1/M_2)\sin\theta)^2]^{1/2}+\cos\theta\}^2}{[1-((M_1/M_2)\sin\theta)^2]^{1/2}}. \tag{4.1}$$

The very strong decrease of the cross section in (4.1) with increasing scattering angle ($1/\sin^4\theta$) and its increase with decreasing projectile energy ($1/E^2$) should be noted. Furthermore, the cross section increases quadratically with the charges Z_1 and Z_2 of the colliding atoms. As discussed above, to obtain a high mass resolution, it is necessary to probe the target with energetic particles scattered as far backwards as is experimentally possible, a requirement which strongly suppresses the cross section. This point is of particular relevance for RBS experiments on carbon-based materials for which the host material is of unusually low mass ($M_2 = 12$). Also of special importance is the fact that (4.1) assumes pure Rutherford scattering, which may not be true for the case of a light element like carbon bombarded by a light energetic projectile (like H). The collision energy in the center of mass system may well exceed the Coulomb barrier in such a case, so that nuclear reactions or resonance scattering will take place, a point which will be discussed below in the section dealing with "Nuclear Reaction Analysis" (Sect. 4.2).

Backward scattered projectiles detected with energies below those corresponding to scattering from the surface of a target of monatomic composition bear information about the depth where the backscattering event took place. The reason for this stems from the fact that for a collision event, which takes place at a distance x from the target surface, the probing particle has to traverse the length x in the solid, while losing energy on its way "in" and, after the scattering event, the probe particle has to find its way out of the target and to reach the detector, losing energy on its way "out". Figure 4.2 defines the geometry and the symbols used to calculate the difference between the energy of a probing particle which has been scattered by the angle θ from a surface atom, kE_0, and the energy, $E_1(x)$, of a particle reaching the detector after suffering a collision at depth x from the target surface:

$$kE_0 - E_1(x) = \left[\frac{k}{\cos\theta_1}\frac{dE}{dx}\bigg|_{\text{in}} - \frac{1}{\cos\theta_2}\frac{dE}{dx}\bigg|_{\text{out}}\right]x. \tag{4.2}$$

The dE/dx values in (4.2) are usually taken at the average energy of the projectile on its way "in" prior to the collision, and on its way "out" after the collision. Equation (4.2) thus converts the energy scale of the detected particle to a depth scale, the highest energy corresponding to scattering from the target surface [$E_1(0) = kE_0$, see (3.4) and (3.5)], and the deeper the scattering, the lower $E_1(x)$. Figure 4.3 shows a schematic spectrum of a light ion beam (He)

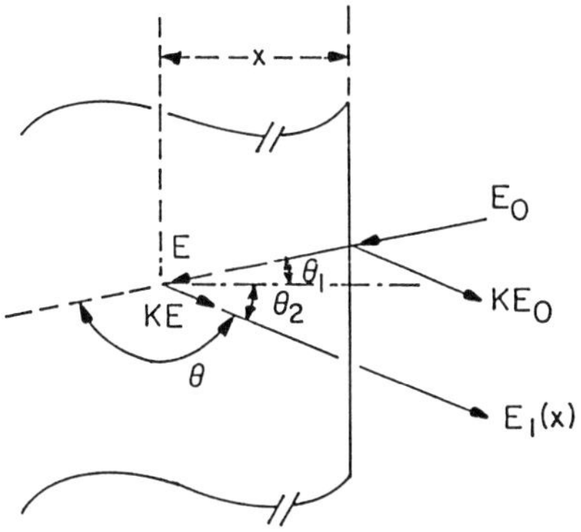

Fig. 4.2. Symbols used in the description of backscattering events in a sample (target) consisting of a monoisotopic element. The angles θ_1 and θ_2 are positive regardless of the side on which they lie with respect to the normal to the sample. The incident beam, the direction of detection and the sample normal are coplanar [4.1]

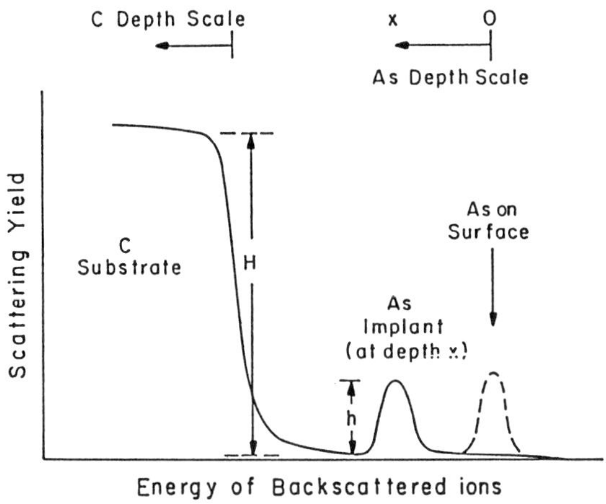

Fig. 4.3. Schematic diagram showing the energy distribution of ions backscattered from a graphite sample (not aligned). The sample was implanted with As atoms to a depth x and was covered with some As atoms on the surface

backscattered from a carbon target implanted with As. Three points should be noted in Fig. 4.3:

1. The continuum spectrum of the C substrate, and its depth scale;
2. The position and width of the peak due to the As implant, which is shifted downwards in energy, and is broadened with respect to the position and width of a thin As layer on top of the C substrate (dotted curve in the figure); and

3. The height of the implanted As peak (h) relative to the near-surface height of the C spectrum (H).

Point (1) is due to the energy dependence of the Rutherford cross section convoluted with the energy loss of the projectile in the target. Point (2) reflects the fact that, because of the higher mass of the arsenic implants, the RBS probe ions backscattered from the arsenic will have higher energies than those backscattered from the carbon atoms, therefore allowing the profile of the arsenic impurity to be measured independently from that of the carbon host atoms. The energy at which the impurity peak appears, relative to the energy that would be expected if this impurity had been on the surface (4.2), yields information on the depth of the implanted impurity, while the width of the peak, corrected for detector resolution, provides information on the diffusion and straggling of the implanted impurity (compare the dotted and solid peaks in Fig. 4.3 which correspond to As on the sample surface and to implanted As, respectively). Point (3) illustrates the fact that the Rutherford backscattering spectrum also gives the number density of a specific atomic species at a depth x from measurements of the height H of the RBS spectrum (Fig. 4.3). The height H is given by

$$H = QNx\sigma(\Omega)\Omega \tag{4.3}$$

Q the total number of particles that hit the target, N the volume density of the target atoms, $\sigma(\Omega)$ the average differential scattering cross section and Ω is the solid angle spanned by the detector. [In writing (4.3) it is assumed that the probe ions are incident along some random (non-channeling) direction with regard to the crystallographic axes.] Usually it is easier to deduce relative concentrations of elements, rather than to use (4.3). Hence the relative heights of the As peak h to the carbon height H would mainly reflect the ratio between the number of As atoms to that of C atoms in the target, corrected for the different cross sections for the two elements [which roughly scale as $(Z_2(\mathrm{As})/Z_2(\mathrm{C}))^2 \sim 30$] and corrected for the different projectile energies prior to the collision event due to the depth of the As implants.

4.2 Nuclear Reaction Analysis

Rutherford backscattering does not always reveal impurities embedded in a host matrix, especially if the mass of the impurities is smaller than the mass of the host atoms. In such cases, ion-induced X-ray and ion-induced nuclear reactions are used as signatures for the presence of the impurities inside the crystal. Here again, the lattice location can be derived from the changes in yield of these processes for random and channeled impingement of the probing beam as will be discussed in Sect. 4.4.

Nuclear reaction analysis (NRA), in contrast to RBS, is performed with the probing beam energy exceeding the energy of the Coulomb barrier for the particular atomic (nuclear isotope) species to be detected. In this case the close encounter event, which is always necessary in IBA to identify the collision, is a nuclear reaction, and the products of this reaction, being charged particles or gamma rays, are the signature that the reaction has taken place. Nuclear reactions, which exhibit sharp resonances at particular energies E_{res} or a sharp threshold at E_{thresh}, are particularly useful as they allow quantitative depth profiling. In such a case, information about the depth x at which the reaction took place inside the target (and hence about the abundance of the particular species at that depth) is obtained through the requirement that for the reaction to occur, the resonance (or threshold) energy must be hit at the depth x. This means that the beam energy at the target surface E_0 must exceed E_{res} by the energy to be lost by the probing projectile during its passage through the target material to the depth x:

$$E_0 - E_{\mathrm{res}}(x) = \int_0^x \frac{dE}{dx'} dx' . \qquad (4.4)$$

By increasing the beam energy above E_{res}, it is thus possible to profile the depth distribution of the desired impurity species. For those cases in which the reaction product is a charged particle, depth information can also be obtained from the shape of the spectrum of the detected particles through their energy loss on their way out of the target from the reaction point at depth x to the detector. The nuclear reaction technique has been described by *Feldman* and *Picraux* [4.4, 6] who also lists the recommended reactions and their specific parameters (energy, reaction products, differential cross sections, etc.). It should be clear that the NRA technique is applicable to only a selected number of cases for which suitable nuclear reactions exist; however for those cases (mostly light nuclei) where the method is applicable, it is very powerful indeed. NRA has been extensively applied among others to the study of carbon-based materials, for the identification of light impurities in natural diamonds [4.7, 8], for the investigation of self diffusion in diamond using ^{13}C as the probe nucleus (^{13}C (p,γ)^{14}N) [4.9], and for the probing of ^{14}N and H (or D) in diamond, graphite and other related materials [4.10-14].

4.3 Particle Induced X-Ray Emission

The close encounter of a projectile ion to a target atom may lead to the ionization of an inner atomic shell of the atom. The hole thus formed will be rapidly filled by an electron from an outer shell, and the energy gained by the electron due to this transition may be emitted as a photon at X-ray energies. The energy of the emitted photon reflects the energy levels of the emitting atom. Accurate identification of atoms in the near-surface region of the target may thus be obtained from such particle-induced X-ray emission (PIXE) experiments [4.4]. However, in contrast with the previously described RBS or NRA methods, which have internal absolute yield and depth calibrations, PIXE is hard to be made quantitative and does not bear accurate depth information due to the complexity of the X-ray formation processes and the attenuation that the photons undergo in the solid on their way to the X-ray detector. Nevertheless, PIXE is well suited for relative measurements, as are needed, for example, for channeling experiments in which yields for aligned (channeled) and random incidence are compared. PIXE also has a very high mass sensitivity due to the availability of high resolution detectors operational in the 2-30 keV X-ray regime. Hence, in contrast to NRA which is useful for only specific cases, PIXE has no limitation on the elements which it can detect, except for the lowest Z atoms, for which the characteristic X-rays are too soft to be conveniently detected. For the particular case of C based materials, this last point is of importance since the C X-rays are among those which are not detectable by common Si(Li) detectors. Hence the high background of signals from the substrate, which often disturbs the detection of low abundance species in the sample, does not exist for the detection of impurities in carbon matrices. Thus, PIXE is an especially useful tool for the compositional characterization of carbon-based materials; PIXE is capable of detecting impurities at concentrations as low as $1{:}10^5$.

4.4 Channeling

When the probing beam is aligned nearly parallel to a close-packed row of atoms in a single crystal target, the particles in the beam will be steered by the potential field of the rows of atoms in the crystal, resulting in an undulatory motion in which the "channeled" ions will not approach the atoms in the row to closer than 0.1-0.2Å (Fig. 4.4). This is called the channeling effect [4.2, 3]. Under this

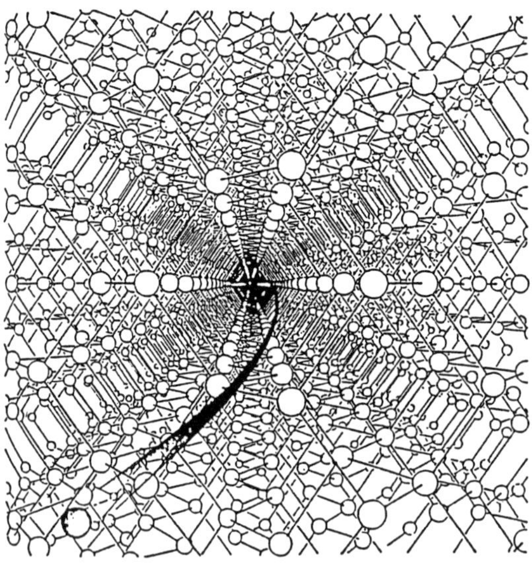

Fig. 4.4. Artists conception of the channeling process on a microscopic scale [4.15]

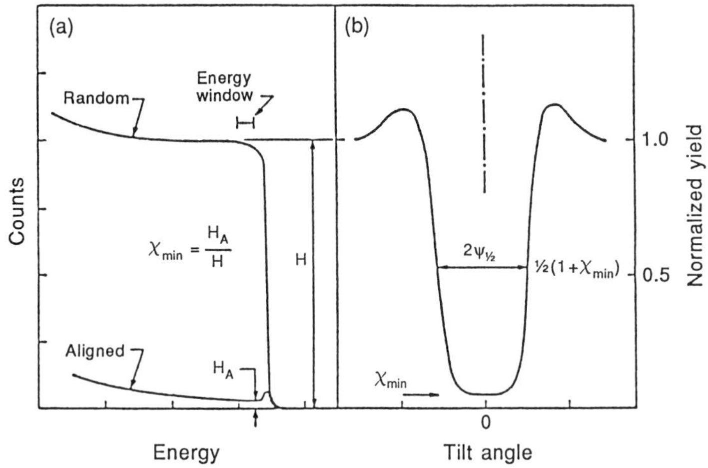

Fig. 4.5. Rutherford backscattering yield in the vicinity of a channeling direction, where (**a**) shows χ_{min}, the minimum RBS yield in the channeling (aligned) direction, in comparison with the RBS yield in a random direction, and (**b**) shows the angular width of the channel, where $\psi_{1/2}$ denotes the angular spread corresponding to the half maximum yield [4.2]

channeling condition, the probability of large angle Rutherford scattering or any other close encounter collision event is greatly reduced. As a consequence, there will be a drastic reduction in the backscattering yields of all processes described above when the probing ion beam is aligned along a major crystallographic direction in a perfect crystal, as compared to the yields measured for the beam incident in a "random" (non-aligned) direction. Figure 4.5a illustrates this for the case of the RBS yield. Even though both "random" and "aligned" spectra were taken for an identical probing ion beam dose (i.e., the same number of incident particles) the number of backscattered events counted in the detector is greatly reduced for the aligned spectrum as a result of the channeling effect. The degree of crystallinity of the target is reflected in this reduction in yield through the "normalized minimum yield" χ_{min}, which is defined as the ratio of the number of backscattered counts in a narrow energy window (set near the crystal surface) between the channeled (aligned) and random spectra ($\chi_{min} \equiv H_a/H$, as shown in Fig. 4.5a). The minimum yield can easily be estimated theoretically. For the case of the minimum permitted distance of approach ρ of the projectile to an atomic row, a crystal having an atomic concentration N and an atomic spacing along the row d exhibits a minimum yield given by

$$\chi_{min}(\rho) = Nd\,\pi\rho^2. \tag{4.5}$$

Interestingly, the quantity χ_{min} depends only on the properties of the crystal, and not on any of the scattering parameters (Z_1, Z_2 or E). More refined calculations, based on computer modeling somewhat modify (4.5); however, these are beyond the scope of this survey (for details see the books by *Feldman* et al. and by *Morgan* [4.2, 3]). The distance of closest approach ρ, which strongly affects χ_{min} as seen from (4.5), is predominantly determined by the thermal vibrations of the atoms in the crystal. These are exceptionally low for diamond (having a very high Debye temperature); hence diamonds of high crystalline perfection exhibit extremely low values of χ_{min}, and channeling in them is particularly useful for crystal quality assessment.

In actual channeling experiments, the crystalline sample is mounted on a goniometer and the close encounter events (like backscattering from the near surface region) are counted as a function of tilt angle (ψ) with respect to the beam for a fixed number of incident particles. The resulting "angular scan" curve, normalized to the counting rate at the random directions, is shown in Fig. 4.5b. The curve is symmetric with respect to the minimum yield and has a width which is characterized by the quantity $\psi_{1/2}$ defined as the halfwidth at half height of the angular scan curve (Fig. 4.5b). A rough estimate of the critical value of the angle ψ_c above which the projectile will "punch through" the atomic string can easily be obtained by equating the transverse energy of the incident particle $E_0\psi_c^2$ to the transverse energy of repulsion from the atomic string at the distance of closest approach $U(\rho)$, yielding

$$\psi_c = [U(\rho)/E_0]^{1/2}. \tag{4.6}$$

More refined calculations lead to results which do not differ much from (4.6), and can be found in the literature [4.2, 3].

The RBS-channeling technique is frequently used to study radiation-induced lattice disorder by measuring the fraction of atom sites for which the channel is blocked. When the probing beam is aligned along a channeling direction of a perfect crystal, a dramatic reduction in the backscattering yield is observed as the channeled ions are steered by the rows of atoms in the crystal, never getting close enough to the atom to undergo a close-encounter interaction of any kind. However, if some portion of the crystal is disordered and lattice atoms are displaced so that they partially block the channels, the ions directed along nominal channeling directions can now have a close collision with these displaced atoms, so that the resulting scattered yield will be increased above that for an undisturbed channel (Fig. 4.6). Furthermore, since the displaced atoms are of equal mass to those of the surrounding lattice, an increase in the RBS yield occurs at an energy in the yield versus energy spectrum corresponding to the depth at which the displaced atoms are located. The increase in the backscattering yield from a given depth will thus depend upon the number of displaced atoms, so the depth (or equivalently, the backscattering energy E_1) dependence of the yield reflects the depth dependence of the displaced atoms. Integrations over areas under the damage peak which correspond to different depth bins (shaded regions in Fig. 4.6) gives (after proper subtraction of the dechanneling background in the RBS spectrum) a measure of the number of displaced atoms within the corresponding depth bins. A variety of techniques have been

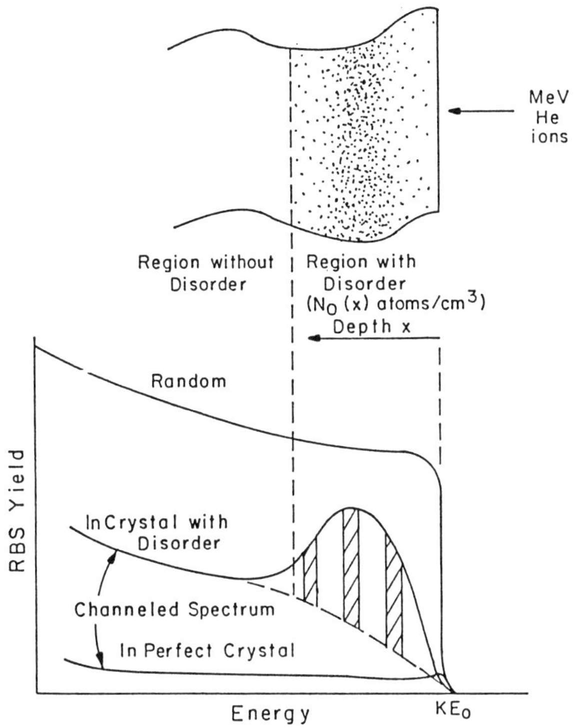

Fig. 4.6. Schematic diagram of the random and aligned RBS spectra for MeV ^{4}He ions incident on a crystal containing disorder in the near surface region. The aligned spectrum for a perfect crystal without disorder is shown for comparison. The difference (*shaded portion*) in the aligned spectra between disordered and perfect crystals can be used to determine the concentration of displaced atoms in the damaged region at various depths. The dechanneling background for the disordered crystal is also indicated

developed to extract the depth distribution of defects from the backscattering spectrum. These techniques are described in detail in [4.1-3].

Another very useful application of the channeling technique is in the determination of the site location of foreign atoms in a host lattice. Since channeled ions cannot approach the rows of atoms which form the channel closer than ~0.1Å, one defines a "forbidden region" along each row of atoms as a cylindrical region with radius ~0.1Å, such that there are no collisions between the channeled particles and atoms located within the forbidden zone. Any atom residing on an atomic row, like an impurity occupying a substitutional site in the crystal, will be "shadowed" and not detectable by any IBA channeling experiment. However, if an impurity is located outside of the forbidden region, it will be detected by the channeled probing beam just like any target particle will be detected by probing particles from a beam which impinges in a random direction. Thus, by comparing the area under the impurity peak observed for channeling with that for random alignments, the fraction of impurities occupying sites outside of the forbidden region of a particular channel (high symmetry crystallographic axis) can be determined. Repeating the procedure for a variety of crystallographic directions allows the identification of the exact lattice location of the impurity atom in the crystal lattice.

The quality of any crystal under study determines the values of $\chi_{\min}$ and $\psi_{1/2}$. For diamonds, a large variation of the channeling parameters can be found in the literature reflecting different degrees of perfection of the crystals. However, due to the high Debye temperature of diamond, rather deep channeling dips can be found in well-polished high quality diamonds. Typical values for channeling of 350 keV protons along the $\langle 111 \rangle$ axis would be $\chi_{\min} \approx 0.05$ and $\psi_{1/2} \approx 0.7°$. These values greatly differ from those found for typical graphite samples (discussed in Sect. 4.7). Hence, the channeling technique offers a much more powerful tool for the determination of lattice damage, and the location of impurity sites in diamond in comparison to graphite. Except for diamond, graphite and some graphite intercalation compounds, the crystalline order in other carbon-based materials is not sufficient for channeling studies.

The RBS channeling technique has been applied extensively to characterize graphite before and after ion implantation. The application of this technique to highly ordered pyrolytic graphite (HOPG) has a number of unusual features because of its unique polycrystallinity that merit special attention. These are discussed below (Sect. 4.7) as background for the application of the RBS channeling technique to ion implanted HOPG, the host material most commonly used for ion implantation studies in graphite. This technique has also been applied in an interesting way to characterize the stoichiometry (Sect. 4.8) and the c-axis alignment (Sect. 4.9) of graphite intercalation compounds, which also have some unique features.

4.5 Elastic Recoil Detection

In contrast to Rutherford backscattering spectroscopy (RBS) in which a backward scattered light projectile is detected following a collision with a heavier target atom, the elastic recoil detection (ERD) technique utilizes the forward momentum imparted to a light target atom by a Rutherford scattering event to profile light elements in the near surface regions of solids. Thus an ERD experiment is, in a sense, complementary to a RBS experiment. The projectiles used are relatively massive and energetic (ranging from MeV ^{4}He ions, mainly used for hydrogen profiling, to heavy ions like ^{37}Cl accelerated to several tens of MeV), and the experimental geometry is such that the beam impinges on the target at an oblique angle with a detector set at a very small forward angle. Those constituents of the target that are counted are those that have experienced an elastic collision with the probing projectiles acquiring a sufficient forward momentum to enable them to leave the target (from the depth where the collision took place) and to reach the detector. Some selection system (like a time-of-flight setup) is used to discriminate between different recoiling species and to determine their energies. Calculations similar to those used in converting the energies of backscattered particles detected in RBS experiments to depth scales are used to obtain depth scales in ERD experiments.

The ERD method is most useful for the detection of light atoms in a solid. It has been extensively used also in carbon-based materials, as for example used by *Boutard* et al. to investigate the hydrogen (or deuterium) content in diamond or in diamond-like films [4.16].

Although, in principle, ERD could also be applied in connection with channeling to obtain information on the lattice location of the light impurities, no report on the successful use of such channeling experiments have been published so far.

4.6 Secondary Ion Mass Spectroscopy

Secondary ion mass spectroscopy (SIMS) is an ion beam probing technique in which species ejected from the target are measured; in this respect SIMS resembles ERD. However, the ejection of constituents of the target is achieved in SIMS by the controlled removal of surface layers due to the sputtering induced by low energy (1-10keV) heavy ions. The sputtered particles are analyzed in extremely sensitive high resolution spectrometers, which have typical $\Delta M/M$ resolutions on the order of 1:10000, and hence can easily resolve isotopes and various molecular species removed from the target.

By following the counting rate of a specific mass as a function of sputter time, depth profiles of rare impurities and of implanted species can be obtained (assuming the sputter rate is known) with sensitivities many orders of magnitude superior to those obtainable by RBS or other profiling techniques. Figure 4.7 shows a typical SIMS depth profile of an impurity implanted un-

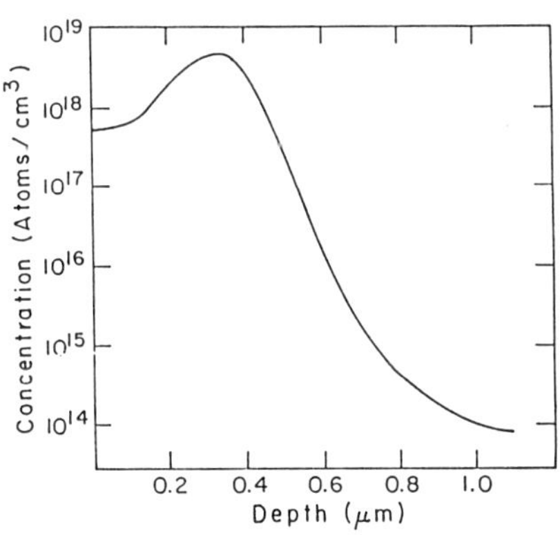

Fig. 4.7. Schematic SIMS profile of impurity atoms implanted into graphite under non-channeling incidence. Note the high sensitivity of SIMS to even low implant concentrations

der non-channeling conditions into carbon. The impurity distribution can be followed by SIMS over many decades of concentration.

By its very nature, it is obvious that SIMS is an ion beam probing technique which is destructive and is not at all sensitive to the crystallinity of the specimen under investigation. Typical crater sizes caused by the SIMS sputtering are less than 1 mm in diameter, while the mass distribution is determined from an area of $\sim$10 μm diameter at the crater center. Thus, SIMS tends to disrupt the surface of the sample, due to the severe bombardment with low energy ions which is inherent to the technique.

For the specific case of probing diamond by SIMS, the destructive nature of the sputtering process is most severe as it tends to graphitize the bombarded area, thus causing very basic alterations to the diamond sample during the analysis process.

4.7 Channeling Studies in Graphite-Based Materials

As discussed in Sect. 4.4, the ion channeling technique has been widely used to study properties of nearly perfect single crystals, of single crystals disordered by ion implantation and of epitaxial layers [4.1]. However, since graphite and related materials exhibit somewhat unusual crystallographic properties, channeling studies in these materials are unique, and are therefore discussed below in some detail. Almost all of the ion channeling studies on ion implanted graphite have been done using highly oriented pyrolytic graphite (HOPG) as the target material [4.17]. HOPG is a polycrystalline material developed and characterized by *Moore* [4.17] with special properties that still permit channeling studies to be carried out. However, channeling in HOPG exhibits a number of different properties from that in single crystal materials. These differences need to be reviewed to provide the background needed for understanding the applications of RBS/channeling studies in ion implanted graphite discussed in Chapter 6.

The structure of HOPG differs from that of other host materials for ion implantation in a number of fundamental ways. Firstly, though aligned along the c-axis, HOPG is a polycrystalline material, with crystallites of $\sim 1\mu$m average in-plane crystallite size that are preferentially oriented to less than 1° from the normal to the c-axis [4.17]. Secondly, HOPG is subject to a considerable amount of lattice disorder that has been classified by *Austerman* [4.18] as: (1) *twist* disorder, in which the basal planes are rotated out of registry with regard to the AB layer stacking, while the graphene planes remain parallel to each other, and (2) *tilt* disorder or mosaic spread, in which the basal planes of different crystallites exhibit a mosaic spread in their c-axis orientations with respect to the mean c-axis direction [4.19, 20]. All these features of the HOPG material have a significant effect on the channeling spectra. Fortunately these features comply with the requirement of an analytical model [4.21, 22] developed for the analysis of epitaxial silicides grown on Si substrates, thereby permitting the use of the model for the analysis of channeling spectra from HOPG.

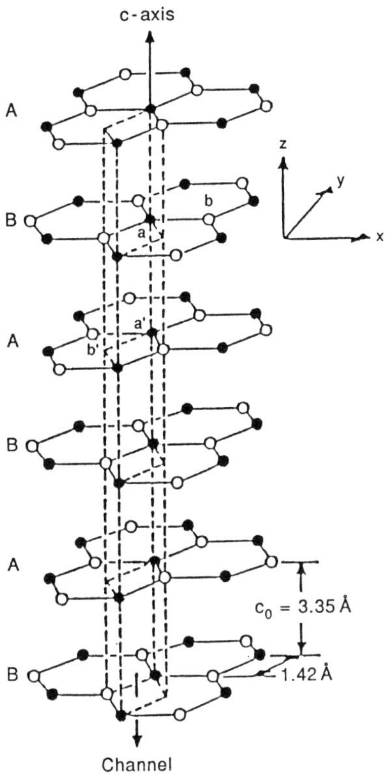

Fig. 4.8. Crystal structure of graphite showing a possible channeling direction along the c-axis. The figure shows that the rows of a and a' atoms (*dark circles*) have an interatomic spacing of c_0 while rows of b and b' atoms (*open circles*) have a spacing of $2c_0$. The AB stacking of the carbon layers is noted [4.23]

Since the in-plane crystalline size is $\geq 1\mu$m in good quality HOPG, the grain size is large enough to exhibit channeling along the c-axis within each crystallite (Fig. 4.8). The distribution of c-axis orientations of different crystallites and the channeling angular yield for each crystallite are assumed to be cylindrically symmetric, with a standard deviation σ and a half-width at half maximum yield $\psi^0_{1/2}$ respectively. The last assumption implies that many crystallites are analyzed simultaneously in a HOPG sample within the 1-mm diameter of the analyzing ion beam. The experimental results for the angular distribution of backscattered $^4He^+$ ions in graphite (HOPG) are shown in Fig. 4.9.

The angular dependence of the backscattering yield in polycrystals is given by the convolution of the angular dependence of the single-crystal yield with the angular distribution of the crystallite orientations. Assuming a Gaussian distribution for the crystallite orientations and a Gaussian functional form for the channeling angular yield in single crystals, *Ishiwara* and *Furukawa* [4.22] showed that the RBS yield for the polycrystalline materials is given by

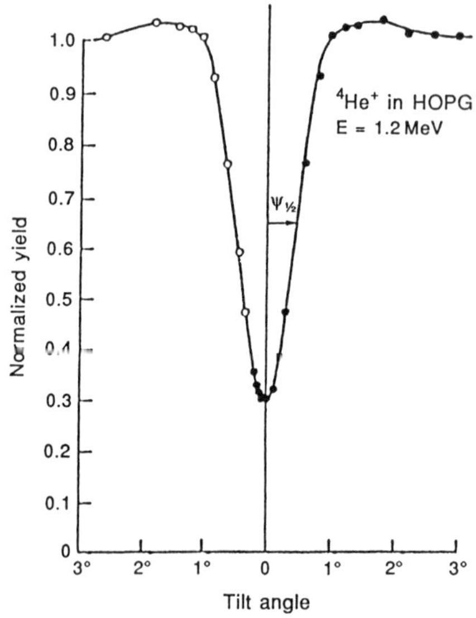

Fig. 4.9. Normalized angular distribution of 1.2 MeV $^4He^+$ ions backscattered from graphite (HOPG). The experimental determination of the small $\psi_{1/2}$ half-width (HWHM) for graphite is indicated. The open points are implied by a fit (*solid curve*) to the measurements (*dark points*). To avoid sample damage, a minimum number of backscattering points are measured [4.23]

$$\chi_{\text{poly}}(\psi, E_0) = 1 - \left\{ \frac{1-\chi_0}{1+(\sigma/\psi^0_{1/2})^2 \ln 2} \right\} \exp\left[-\frac{\psi^2}{\sigma^2 + (\psi^0_{1/2})^2 \ln 2} \right], \tag{4.7}$$

where $\chi_{\min} = \chi_{\text{poly}}(0, E_0)$ is the minimum yield for the polycrystalline material, E_0 is the incident energy, ψ is the tilt angle of the beam from the most preferred channeling axis in the polycrystal, σ is the standard deviation of the spread in c-axis crystallite orientations, and $\psi^0_{1/2}$ and χ_0 are the critical angle and minimum yield for channeling in the single crystal, respectively. Combining (4.7) with an expression for the energy dependence of the backscattering yield [4.1, 2] gives expressions for the minimum yield and for the half width in the analysis of RBS spectra for HOPG:

$$\chi_{\min} = \frac{\chi_0 + (\sigma/\psi^0_{1/2})^2 \ln 2}{1 + (\sigma/\psi^0_{1/2})^2 \ln 2}, \tag{4.8}$$

and

$$(\psi_{1/2})^2 = \sigma^2 \ln 2 + (\psi^0_{1/2})^2, \tag{4.9}$$

in which $\chi_0, \psi^0_{1/2}$, and $\chi_{\min}, \psi_{1/2}$ are the minimum yield and half width at half maximum angular width for single-crystal and polycrystalline graphite, respectively. By measuring the angular and energy dependences of the backscattering yield, it has been possible to deduce the minimum yield χ_0 and the critical angle $\psi^0_{1/2}$ for single-crystal graphite (which cannot be determined directly because of

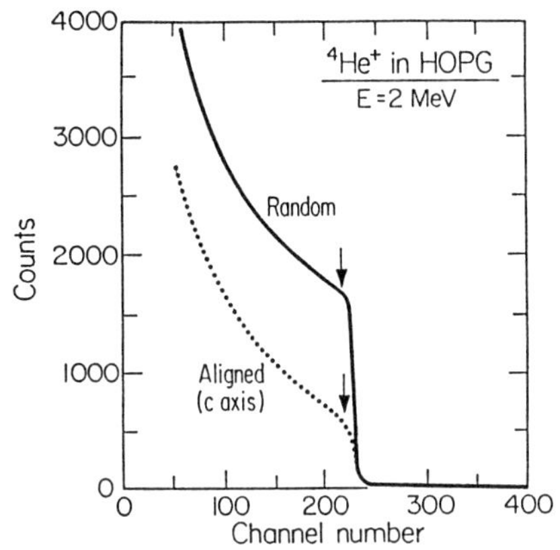

Fig. 4.10. Energy spectra for 2-MeV helium ions backscattered from an HOPG sample in random and aligned (channeling along the c-axis) directions. The arrows indicate the channel number (energy) at which $\chi_{\min}$ is determined from the ratio of the aligned yield to the random yield [4.23]

the difficulty in obtaining appropriately large single-crystals) and the standard deviation of the spread of crystallite c-axis orientations for specific HOPG samples. Specific RBS results for HOPG have been obtained by *Iwata* et al. [4.24] and by *Elman* et al. [4.23]. *Elman* et al., using a 2 MeV He^+ probe, combined their results with those of *Iwata* et al. to obtain $\chi_0 = 6\%$, and $\sigma = 0.24° \pm 0.04°$ using methods previously developed by *Barrett* [4.25] and *Gemmell* and *Mikkelson* [4.26] for other classes of materials.

Typical aligned and random energy spectra by *Elman* et al. of the yield of 2-MeV$^4He^+$ ions backscattered from HOPG [4.23] are shown in Fig. 4.10. The aligned spectrum corresponds to channeling, where the analyzing beam impinges on the sample parallel to the highly preferred c-axis orientation, while for the random spectrum, the $^4He^+$ beam is tilted $\geq 10°$ away from the channeling direction. Two characteristic features of HOPG (that differ from the channeling in a single-crystal) are observed in Fig. 4.10: the aligned spectrum does not exhibit a surface peak and the value of the minimum yield $\chi_{\min}$ is rather high ($\sim 30\%$). Both effects are due to the polycrystalline nature of HOPG discussed above.

An unusual increase in $\psi_{1/2}$ is observed with increasing depth (Fig. 4.11), and is associated with the special type of polycrystalline morphology of HOPG. This result is in contrast to the usual decrease of $\psi_{1/2}$ with depth found in typical channeling studies of single-crystal semiconductors and metals [4.2, 27]. In a typical single-crystal material, the critical angle for channeling decreases with depth because those channeled ions entering the crystal at angles larger than $|\psi_{1/2}|$ will acquire enough transverse momentum to be dechanneled at a shallower depth than the ions that enter the crystal close to the axial direction. Those ions entering the crystal in a nonchanneling direction have a negligi-

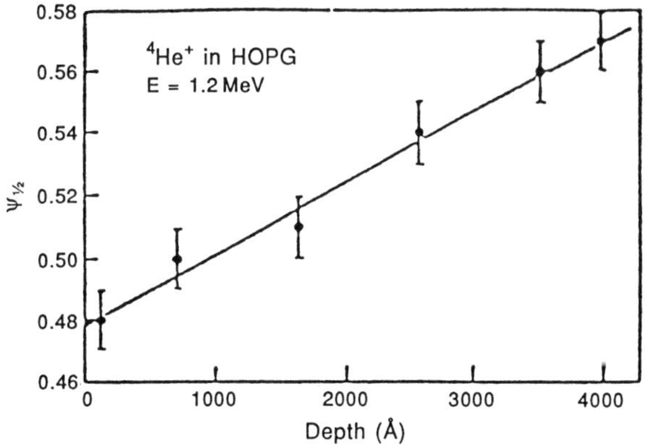

Fig. 4.11. Depth dependence of $\psi_{1/2}$ (in degrees) for channeling of 1.2-MeV$^4He^+$ ions in HOPG along the c-axis. Note the anomalous increase in $\psi_{1/2}$ with increasing depth reported by *Elman* et al. [4.23]

ble probability to become channeled inside the crystal, even if their angle of incidence is very close to $\psi_{1/2}$.

The behavior of the dechanneling with depth is quite different in a polycrystalline material like HOPG, having a spread of channeling directions which can be approximated by a Gaussian distribution that does not change with depth. Those ions that acquire enough transverse momentum to be dechanneled have a high probability to find, deeper in the crystal, a crystallite suitably oriented for channeling. As a result, the critical angle can remain nearly unchanged with increasing depth. Furthermore, a collimated beam of ions, entering the crystal in a nonchanneling direction will be broadened because of multiple scattering. If the angle of incidence is close to $\psi_{1/2}$, an appreciable fraction of the incoming ions will be able to find an appropriate crystallite with a favorable c-axis orientation for channeling deeper in the crystal. The spread of the incoming beam increases with depth and consequently the probability that these ions will find properly oriented crystallites increases with depth, thus accounting for the observed increase in the critical angle for channeling with increasing depth.

4.8 Stoichiometric Characterization of GICs by RBS

An important application of the RBS technique to graphite-related materials is the determination of the stoichiometry of graphite intercalation compounds (GICs) (Sect. 2.1.5). This stoichiometric information is of importance to the characterization of GICs, on one hand, and to the characterization on ion implanted GICs, on the other hand. Thus far, no work has been published on ion implantation into GICs.

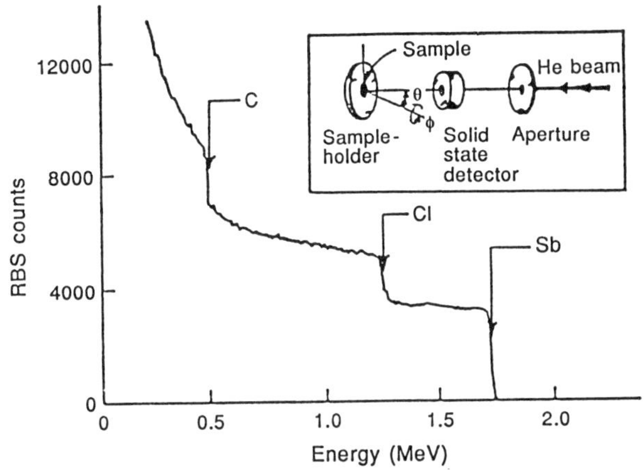

Fig. 4.12. Typical RBS spectrum of a stage 3 $SbCl_5$–GIC sample using 2 MeV $^4He^+$ ions and an analyzing angle of 175°. The number of incoming ions was typically preset at $\sim 1\mu C$ of charge. The inset shows the experimental geometry used for the RBS experiments [4.30]

Intercalate stoichiometry is a key issue for the determination of the structural phases and phase transitions in GICs. Analysis of the energy distribution of the backscattered ions provides detailed stoichiometric information as a function of depth from the surface. This information is averaged over an area that approximates the diameter of the $^4He^+$ ion beam, which is typically 1 mm. The lateral distribution of the intercalate species can also be monitored by taking spectra at different positions on the sample. Such information is not yet available using other nondestructive techniques. The use of RBS for stoichiometric studies in a particular case is described in some detail in this section.

A detailed study using the RBS channeling technique to determine stoichiometric ratios between the elemental constituents of a graphite intercalation compound has been carried out for the $SbCl_5$–GIC system, formed by intercalation of $SbCl_5$ into graphite. From studies of the $SbCl_5$–GIC system by *Salamanca-Riba* et al. and by *Boolchand* et al. using other techniques, there is evidence for a deviation from the stoichiometric Cl:Sb ratio of 5:1 on a microscopic scale in the intercalation compound [4.28, 29]. The depth sensitivity of the RBS technique makes it possible to determine the stoichiometric ratios of the constituents of the GIC. It was found that the stoichiometry near the surface of as-prepared intercalated graphite samples may differ significantly from the Cl:Sb ratio of 5:1 and also from the Cl:Sb ratio in the bulk.

To carry out this study, RBS spectra of $SbCl_5$–GICs with various stages (n = 2, 3, 4 and 6) [4.30] were taken using 2MeV$^4He^+$ ions both before and after cleaving. A typical RBS spectrum from a cleaved stage 3 $SbCl_5$–GIC sample shows contributions from specific atomic species, as indicated in Fig. 4.12. Specifically, the three sharp steps at energies 1.755, 1.274, and 0.502 MeV correspond, respectively, to the energies of backscattered ions from ^{121}Sb, ^{36}Cl, and ^{12}C atoms on the sample surface.

The RBS spectrum of Fig. 4.12 corresponds to a well-staged layered material, but because of the small thickness of the *c*-axis unit cell (between 12.7 Å and 26.3 Å) and because of the poor depth resolution for the case of carbon (~ 450Å), the $^4He^+$ probes in the RBS experiment could not resolve the differences in the chemical species associated with individual layers of the graphite intercalation compounds. The spectrum of Fig. 4.12 is thus indistinguishable from that for a homogeneous multi-elemental sample with a stoichiometry $C_{40.6}SbCl_{4.6}$.

To extract the stoichiometry quantitatively from the heights of the steps of the RBS spectrum (Fig. 4.12), the relative atomic concentration for each of the elements was obtained by *Elman* et al. from the measured relative heights of the RBS signal at the surface edge taking into account the differential scattering cross sections and stopping cross section factor for each element [4.30]. The RBS spectral edge heights and lineshapes for carbon and chlorine were observed to vary little between different freshly prepared samples of similar stage [4.30]. No change in the stoichiometry (within experimental error) was found in the lateral directions for cleaved and uncleaved samples. By taking spectra on samples that were cleaved, it was found that all three elements are homogeneously distributed in depth, all the way from the surface of the sample to a depth on the order of several microns.

RBS spectra have also been reported by *Elman* et al. for stoichiometric determinations in donor GICs [4.30], specifically a KHg–GIC. From the depth dependence of the stoichiometry as a function of lateral position (on a 1.5 × 1.5mm^2 sample), it was found that regions near the edges showed a uniform depth distribution of Hg and K, while the central region showed a decrease in Hg and K with depth. In both cases, however, the stoichiometry at the surface was the same, consistent with the idea that intercalation occurs as the intercalate penetrates the sample from *a*-face edges with the intercalation proceeding from the outermost planes toward the bulk.

Thus RBS spectra provide a powerful tool for characterizing the stoichiometry of layered compounds such as graphite intercalation compounds, yielding information not readily available by use of other techniques.

4.9 Ion Channeling in GICs

An even more challenging application of ion beam techniques to graphite-related materials is the demonstration by *Salamanca-Riba* et al. of ion channeling in graphite intercalation compounds (GICs) [4.31]. Channeling experiments performed as a function of temperature provided vital information on a variety of structural properties like the stacking of graphite planes, the commensurability of the intercalate layer relative to the graphite layers and atomic rearrangement transitions. The study of ion channeling in GICs was pioneered by the MIT group, and the literature here is confined to a single published work [4.31] on a stage 4 HOPG-based K–GIC. In this work RBS-channeling analysis was carried out using a beam of 500 keV protons in order to enhance the backscattering yield of C atoms relative to K atoms and thereby to facilitate the alignment of the sample with its *c*-axis parallel to the probing beam.

Since the steering of the channeled particles is established by the host crystal lattice, the observation of the channeling effect in the K-intercalated HOPG sample indicates that the stacking is preserved between the graphite layers above and below the intercalate layer. Furthermore, the similar values found for the axial half angle $\psi_{1/2}$ for the carbon signal, in both pristine HOPG and the K–GIC indicates that the host graphite lattice is not greatly changed by intercalation. The observed increase in the minimum yield $\chi_{\min}$ of the C signal from the K–GIC with respect to HOPG can be attributed to the reduced crystalline order of the near-surface region of the intercalated sample. This assumption is supported by the observation of a weak dependence of the normalized yield vs depth, indicating rather perfect long range crystalline coherence [4.31].

An important result of the study by *Salamanca-Riba* et al. [4.31] was the observation of a reduced yield for the K signal in the K–GIC for beam alignment in the channeling direction, when the measurement was performed at room temperature, indicating that at ~300 K a substantial fraction of the K atoms are located along the rows of carbon atoms. Such a position is consistent with K atoms sitting below or above the center of the hexagon of carbon atoms of the adjacent graphite planes. The fraction of K atoms located in this position can be determined using the relation [4.1]

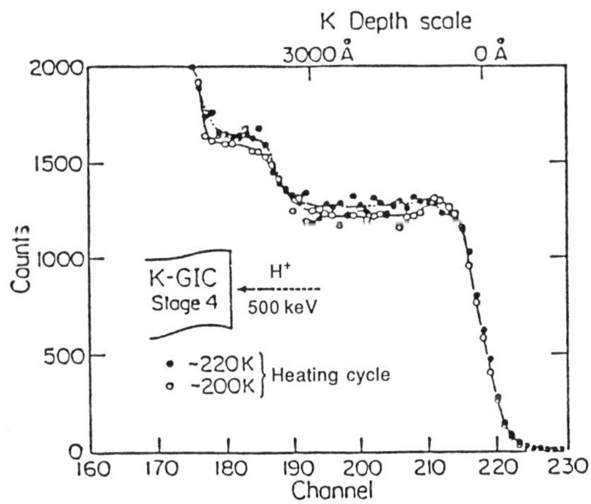

Fig. 4.13. RBS spectrum (random) corresponding to K atoms in a stage 4 K–GIC sample. The RBS probe ions are H^+ ions at 500 keV. The spectrum taken at 220 K (*full circles*) shows a higher K concentration than the spectrum 200 K (*open circles*) [4.31]

$$f = \frac{1 - \chi_{\text{impurity}}}{1 - \chi_{\text{host}}} \tag{4.10}$$

which yielded f=0.30. The location of other K atoms could not be determined precisely.

The angular dependence of the yield of the carbon signal in the K–GIC was nearly unchanged between measurements at 100 K and room temperature. Also no reduction in the angular dependence of the yield for the K signal was observed when scanning through the channeling direction for the host material. This implies that the K atoms are randomly located with respect to the rows of carbon atoms, and the crystalline structure of the HOPG host is not changed upon intercalation. Temperature-dependent RBS studies gave evidence for structural phase changes at temperatures similar to those reported in the review article by *Dresselhaus* and *Dresselhaus* using other techniques [4.32].

The unique information obtained from this application of the RBS-channeling technique to GICs (Fig. 4.13) suggests that further investigations should be carried out to study the role of the in-plane intercalate ordering on the long-range ordering of the graphite skeleton in GICs. Unfortunately, this work on GICs was never continued beyond the initial study.

5. Other Characterization Techniques

Although ion beam techniques provide the most powerful methods for characterizing ion-implanted carbon-based materials, a number of other techniques have also yielded important characterization information. In this chapter we briefly review the most important of these techniques, indicating how they are used specifically for the characterization of graphite, diamond and related materials.

5.1 Raman Spectroscopy

Raman scattering, the inelastic scattering of photons by the fundamental excitations of the solid, provides a general spectroscopic technique for the non-destructive identification of the symmetries of molecular structure and bonding, and the energies associated with the fundamental excitations of the solid. Furthermore, measurement of the intensity of symmetry-forbidden Raman lines provides information on the degree of disorder of the system.

The large differences in the Raman spectra of the various forms of carbon covered in this article make the Raman technique particularly attractive for distinguishing one form of carbon from another. In addition, Raman spectroscopy is a particularly useful characterization technique for carbon-based materials insofar as specific defect-induced features in the Raman spectra are observed, and these features can be directly related to the implantation process itself. Thus, Raman spectroscopy has become a widely used technique for the characterization of ion-implanted carbon-based materials, both for graphite and graphite-related materials [5.1-6] and for diamond and diamond-related materials [5.3]. Raman spectroscopy also provides from the work of *Wada* et al. a very sensitive probe of small amounts of sp^2 bonding in diamond films because of the ~50 times larger Raman cross section for the sp^2 graphite vibrations relative to the sp^3 diamond-like vibrations [5.3], hence its usefulness for the evaluation of diamond films. In this section, a brief review is given of the use of Raman spectroscopy for the characterization of ordered and disordered graphite-related materials, followed by a brief review of the characterization of diamond and diamond-like films.

Because of the long wavelength of light compared to Brillouin zone dimensions, only the zone center phonons are observed in the first-order Raman spectra for single crystal graphite and diamond. Thus for the case of ideal single crystal graphite it is only the E_{2g_2} Raman-allowed 1582 cm^{-1} mode at the zone

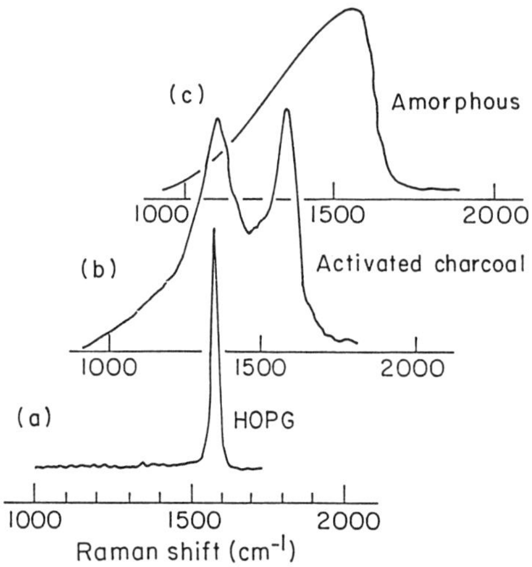

Fig. 5.1. Room temperature Raman spectra for various carbon materials: (**a**) HOPG, (**b**) the D and G lines at 1360 and 1580 cm^{-1} for activated charcoal, (**c**) the broad spectrum for amorphous carbon [5.5]

center ($k = 0$) that is observed in the Raman spectrum at optical frequencies. (There is also a low frequency E_{2g_1} Raman-allowed mode at 42 cm^{-1} but this mode has not been used for characterization purposes.) Disorder associated with the implantation-induced lattice damage causes a broadening of the E_{2g_2} Raman-allowed line and breaks down the $k = 0$ selection rule, giving rise to a disorder-induced (D) line with a peak near 1360 cm^{-1} (Fig. 5.1), corresponding to the maximum in the phonon density of states (Fig. 2.10) [5.7, 8]. The ratio of the integrated intensity of the disorder-induced (D) line to the Raman-allowed (G) line $R = I_D/I_G$ provides a sensitive characterization of the implantation-induced disorder. By calibration against the in-plane crystallite size L_a determined from X-ray diffraction linewidths, *Tuinstra* and *Koenig* [5.4] have shown that $R = I_D/I_G$ depends inversely on L_a, an important characterization parameter for disordered graphite. Recently, *Knight* and *White* [5.9] have shown (Fig. 5.2) the dependence of I_D/I_G on L_a to hold over the extended range $25 < L_a < 3000$Å and for laser wavelengths of 4880Å and 5145Å [5.9]. Their results are shown in Fig. 5.2 as a log-log plot of L_a vs R. As the E_{2g_2} mode is an in-plane vibration, it is predominately sensitive to L_a, the in-plane crystallite size. Since there are no Raman-active c-axis modes, Raman spectroscopy does not provide direct information on L_c, the c-axis crystallite size. The relation between L_a and R in Fig. 5.2 is also applicable to graphite-related materials such as glassy carbon and carbon films. The Raman technique becomes insensitive for the determination of L_a for $L_a \leq 25$Å.

In general, when the exciting photon frequency does not correspond to an interband transition, the Raman effect is a second order virtual process. In this process the incident photon excites the system to a virtual electronic state, and

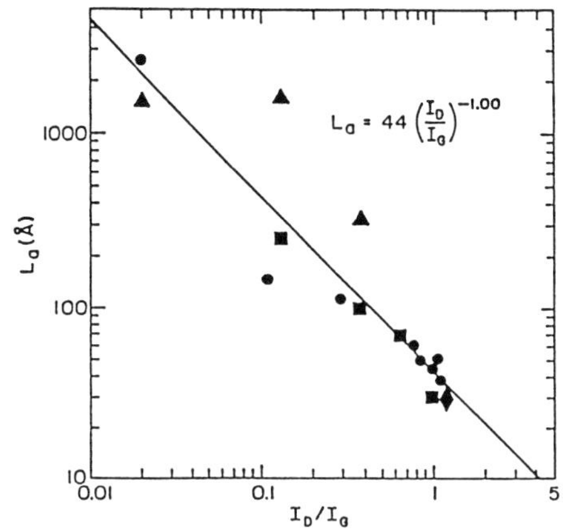

Fig. 5.2. Literature data relating L_a to the Raman intensity ratio I_D/I_G for disordered graphites. The functional form of L_a as a function of I_D/I_G is indicated on the figure [5.9]

on de-excitation the emitted photon can be upshifted by phonon absorption or downshifted by phonon emission, thereby constituting the Raman lines in the emission spectrum. The intensity of the Raman lines is greatly increased when the laser frequency approaches an interband transition frequency. This effect is called resonant enhancement. Since the Raman cross section for each Raman line is frequency dependent, the relative intensities of the various spectral features depend on the laser frequency.

Resonant enhancement effects are operative outside the range of laser wavelengths between 4880 Å and 5145 Å. The resonant Raman phenomenon affects the relative intensities I_D and I_G differently from what is shown in Fig. 5.2, thereby causing the results of Fig. 5.2 to break down outside this range of laser wavelengths. In such cases a direct calibration of I_D/I_G vs crystallite size L_a (from X-ray data) is needed to obtain an I_D/I_G vs L_a plot.

Figure 5.1 shows the Raman spectra for highly oriented pyrolytic graphite (a single line at 1582 cm^{-1}), while activated charcoal shows a broadened G line near 1580 cm^{-1} and a second broadened D line near 1360 cm^{-1}. As the disorder increases these lines eventually merge to yield the spectrum shown in Fig. 5.1 for an amorphous carbon film. Analysis of the spectra for a disordered carbon (such as an ion-implanted graphite sample) is carried out by performing a deconvolution of the Raman spectra into two Lorentzian lines (the D line and the G line) to obtain peak frequencies, linewidths and integrated intensities. Those are then compared and their dependences on ion species, energy, fluence and perhaps other implantation parameters are deduced.

Other features in the first-order Raman spectra that have been used to characterize the implantation-induced disorder in ion-implanted graphite include the intensity of the disorder-induced feature near 1620 cm^{-1} (associated with the maximum frequency of the graphite phonon modes as shown in Fig. 2.10), the

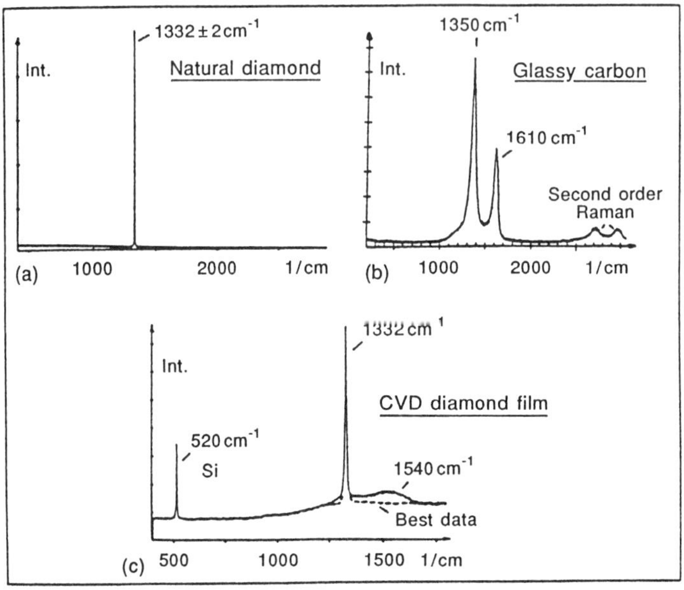

Fig. 5.3. Comparison of the Raman spectra of (**a**) natural diamond, (**b**) glassy carbon and (**c**) a microwave plasma deposited CVD diamond film on a silicon substrate [5.11]

linewidths of the 1360 cm^{-1}, 1582 cm^{-1} and 1620 cm^{-1} Raman-lines, and their frequency shifts. In addition, features in the second-order Raman spectrum have been used to characterize implantation-induced disorder, including the broadening of the Raman-allowed feature near 2730 cm^{-1} and the appearance of a disorder-induced feature at 2970 cm^{-1}; here, the intensities, linewidths and frequency shifts of the 2730 cm^{-1} and 2970 cm^{-1} features are used to characterize the ion-implanted samples [5.5, 6]. Both first- and second-order Raman features have been used to characterize the regrowth of the disordered graphite under various annealing conditions.

With regard to observation by *Solin* et al. of the Raman effect in diamond, which has a space group $P6_3/mmc$, a single sharp peak is observed in the one-phonon spectrum at 1332 cm^{-1} [5.10] (Fig. 5.3). Disordered diamond is expected to show a disorder-induced peak in the 1200–1300 cm^{-1} range [5.12]. Increasing the laser excitation energy to above 3.0 eV reduces the luminescent background, allowing clear observation of the second-order diamond Raman lines, as shown by the work on *Wagner* et al. [5.13]. No resonant enhancement of the diamond lines nor shift in frequency has been observed for laser excitation energies from 2.4 to 4.8 eV [5.13]. Because of the very much larger Raman cross section for graphite sp^2 bonds [5×10^{-5} $cm^{-1}sr^{-1}$] relative to diamond sp^3 bonds [9×10^{-7} $cm^{-1}sr^{-1}$], it is very difficult to observe any spectral Raman features associated with a small amount of sp^3 bonding in the presence of sp^2 bonding [5.3]. By the same token, the high sensitivity of the Raman spectra to

sp^2 bonding allows sensitive detection of small concentrations of sp^2 bonding in a diamond film.

It should be mentioned here that Raman spectroscopy is widely used to characterize diamond films which are currently grown by a wide variety of deposition techniques [5.14, 15]. The occurrence of the sharp line at 1332 cm^{-1} in the Raman spectra of the films is indicative of the growth of diamond, whereas the appearance of features in the Raman spectra centered around 1540 cm^{-1} proves that some graphite is present in the film. Figure 5.3 shows typical Raman spectra of natural diamond, of glassy carbon and of a CVD deposited diamond film, demonstrating the above points [5.11].

The published Raman spectra for various amorphous hydrogenated (a–C:H), diamond-like carbon (DLC) films show many variations from one spectrum to another with regard to peak positions, linewidths, and lineshapes. This is due to differences in the contributions from diamond-like structures, graphite-like structures, disordered carbon, and the amount of hydrogen in the films, arising from differences in preparation conditions [5.3, 9, 16, 17]. A broad peak near 1540 cm^{-1} is seen in the DLC Raman spectra in addition to the broadened diamond line at 1332 cm^{-1}. For DLC films with a higher concentration of sp^2 bonds, the structure near 1540 cm^{-1} is greatly enhanced relative to the diamond peak and a shoulder appears near 1350 cm^{-1} associated with the broad line around 1540 cm^{-1}; such spectra are reminiscent of the Raman spectra for disordered graphite (Fig. 5.1).

Since the DLC films are normally deposited on substrates, the Raman spectrum from the substrate (e.g., Si) is often seen. In addition, the presence of hydrogen in the films alters the C–C distances, which results in large shifts (tens of cm^{-1}) of the graphitic Raman line.

Resonant Raman effects are also observed for a–C:H films [5.18, 19], with huge shifts in the peak position of the Raman line (e.g., Raman peaks at $\sim$1500 cm^{-1} for $E_L = 2.18$ eV and $\sim$1600 cm^{-1} for $E_L = 3.54$ eV). An enhancement of the D line relative to the G line is also observed for low laser energies E_L. For all these reasons, it is necessary to carefully specify the Raman spectra for each DLC film prior to ion implantation, including the specification of the film preparation conditions that were used and the parameters of the Raman measurement.

Some practical considerations in applying the Raman technique to the characterization of ion-implanted materials include proper surface preparation to obtain reliable light scattering results. For graphite, good optical surfaces can be prepared by cleaving (e.g., peeling with Scotch tape) and polishing for the case of diamond and glassy carbon. Attention also has to be paid to matching the optical skin depth to the peak of the damage profile. For example, implantation with light mass ions, or ions at high energies (MeV) often results in lattice damage largely confined to a region well below the surface, such that only a few ions come to rest and little lattice damage occurs within the optical skin depth for visible light.

5.2 Other Optical and Magneto-Optical Techniques

Optical and infrared techniques are generally used to probe the phonon spectra, and to provide information on the free carrier density and on interband transitions in solids. The degree of structural disorder can be characterized by the intensity of symmetry-forbidden features in the infrared spectra. The presence of specific chemical bonds can be monitored through the intensity of their characteristic spectral lines.

The specific optical techniques that are pertinent to the various carbon-based materials differ greatly because of their different optical properties. For example, diamonds which are transparent to visible light are conveniently studied by optical transmission and absorption techniques, whereas the opaque graphitic materials must be examined by IR light in the reflectivity configuration. Infrared reflectivity techniques are especially appropriate for the characterization of ion-implanted materials for those cases in which the optical penetration depth and the projected range of the implanted ions are of comparable magnitude. Infrared spectroscopy has also been widely used to characterize as-grown amorphous hydrogenated (diamond-like) films. For these films, the extensive work of *Dischler* and coworkers [5.20] has provided a detailed listing of all infrared modes found in these materials and their identification with various C–C and C–H bonding configurations. Infrared spectroscopy has nevertheless not yet been used extensively to study implantation-induced changes in a–C:H.

Magneto-optical techniques have also been used to characterize ion-implanted carbon-based materials, specifically graphite at low implantation doses. Because of the high optical absorption of graphite, these experiments must be done in reflection. In the magnetoreflection technique, resonant interband transitions between Landau levels, arising from the high density of states at the extrema of each magnetic subband, are studied as a function of photon energy and magnetic field. By fitting the magnetoreflection spectra of resonant interband transitions to a model for the electronic structure, many of the parameters of a symmetry-based band structure model can be sensitively determined. The effect of ion implantation in modifying the form of the band structure model and in modifying the values of the band parameters can thus be sensitively studied as shown by the work of *Elman* et al. [5.21].

One of the most important issues in characterizing disordered carbons, particularly diamond-like carbon films, is the ratio of the number of sp^2 to sp^3 bonds. In this connection *Savvides* has shown [5.22, 23] that, by measuring the optical transmission and reflection and deducing from the data the real and imaginary parts of the dielectric constant [$\varepsilon_1(\omega)$ and $\varepsilon_2(\omega)$], an estimate for the sp^2/sp^3 ratio could be obtained.

Since diamond is a widegap semiconductor, optical studies of diamond have focused on the characterization of features in the optical spectra associated with defects naturally occurring in diamond. Both optical transmission and luminescence techniques have been successfully employed to sensitively probe defects in bulk diamond, in a–C:H (diamond-like) films and in diamond films.

These powerful techniques have, however, not been applied to the study of implantation-induced defects in diamond, nor in related materials, despite the wide knowledge which exists about the optical properties of these systems prior to ion implantation.

Optical reflectivity studies have not yet been used to characterize implantation-induced modifications of the electronic structure of opaque carbon-based materials, largely because the optical structure due to electronic transitions tends to be quite broad, so that the changes induced by ion implantation can not be measured as sensitively as by other techniques. Although the selection of the wavelength and optical method so as to best probe the volume affected by the ion-implantation is not the only consideration; the sensitivity of a characterization probe to the types of ion beam modifications induced in the material is another important factor in determining the utility of a specific probe.

Infrared spectroscopy of phonon modes could provide a sensitive probe for deep implants (low mass and/or high energy ions) into light absorbing carbons, because the lower frequencies of the electromagnetic radiation used to investigate infrared-active modes (≈ 1000 cm^{-1}), have a deeper electromagnetic skin depth, in comparison to the laser frequencies usually used in Raman spectroscopy. For example, in the case of graphite-based materials, the introduction of sufficient lattice disorder decouples adjacent graphene layers thus suppressing the infrared-active in-plane mode at 1590 cm^{-1} and the c-axis mode at 880 cm^{-1}. Both of these modes depend on the presence of long-range c-axis coherence for their observation. With regard to single-crystal diamond, there are no infrared-active phonon modes. However, in the presence of lattice disorder, the selection rules for infrared transitions break down, so that infrared spectroscopy can be used to characterize specific otherwise symmetry-forbidden modes in the damaged lattice; this type of experiment has however not yet been done.

Magneto-optics is a more specialized technique that has been used by *Elman* et al. to characterize ion-implanted graphite (Sect. 6.4), because of its high sensitivity to the electronic structure [5.21]. This technique could perhaps also yield information on ion-implanted diamond; however, to the best of our knowledge, this technique has not been applied so far to diamond. In the case of graphite, the pertinent photon energy range for the magnetoreflection experiment is in the infrared, where the skin depth is several thousand Å and is usually larger than the region that is modified by the ion beam. Furthermore, the magnetoreflection technique is only sensitive to the spatial regions where the condition $\omega_c\tau > 1$ is satisfied (i.e., the region where the electrons can complete a cyclotron orbit before being scattered). This condition ($\omega_c\tau > 1$) is generally not satisfied in the regions where heavy lattice damage has occurred. Despite these limitations, the magnetoreflection technique has nevertheless provided valuable information on the effect of ion implantation on modifying the electronic structure of graphite in the low fluence limit [5.21].

5.3 Electron Microscopies and Spectroscopies

Since the probing depth for low energy electrons (∼ 1keV) is just a few lattice constants, high energy electrons (e.g., ∼100 keV as are provided by a transmission electron microscope) are commonly used for the characterization of ion implanted materials. Selected area diffraction patterns are often used on relatively well-ordered materials, from which dark field images can be produced and can be used to obtain information about defects associated with crystallites oriented along specific crystallographic directions.

High resolution transmission electron microscopy (TEM) has been extensively used by *Salamanca-Riba* et al. to characterize the microstructure of ion-implanted graphite, with particular attention given to the implantation-induced lattice damage and the subsequent regrowth upon annealing [5.24]. By implanting heavy ions at low energies (such as ^{209}Bi at 30 keV) into graphite (HOPG), the lattice damage can be conveniently confined to a shallow penetration depth ($R_p \sim 150$Å) so that it can be well matched to the sensitivity of a 200 keV high-resolution transmission electron microscope. Bright field images taken with such an instrument have provided a fruitful method for characterizing the lattice damage (Sect. 6.2). Lattice fringe images based on selected area diffraction patterns have been used to yield values of the in-plane and c-axis crystallite sizes (L_a and L_c). In addition to these measurements of the parallel lengths of the lattice fringe segments, L_a and L_c measurements have been made using the longitudinal and transverse widths of the (002) electron diffraction lines. Since the TEM provides two-dimensional information, it is necessary to make a correction for the projection effect to the measured L_a and L_c values [5.24]. By carrying out L_a and L_c measurements as a function of temperature, activation energies for graphite crystallization can be obtained. Through the measurements of the time dependence of L_a and L_c at constant temperature, information on the kinetic mechanisms can be extracted [5.24]. Optical diffractograms taken from the lattice fringe images have been used, with proper calibration, to yield information about interlayer spacings in graphite [5.24].

Scanning electron microscopy (SEM), a standard characterization technique for probing the surface morphology, has been used to characterize all types of carbon-based materials, and to study the effect of ion implantation on the surface morphology. Scanning transmission electron microscopy (STEM) has been applied to study the surface morphology under higher resolution.

The reflection high energy electron diffraction (RHEED) patterns for less ordered materials can be taken, using for example 100 keV electrons which provide a sensitivity to the microstructure within about 200Å from the surface [5.25]. This method, which has effectively been employed by *Prawer* et al., mainly reveals the presence and nature of small (∼100 Å) crystallites. The RHEED technique is therefore particularly useful to probe the formation of crystallites of new phases (e.g., graphite) in damaged diamond.

Auger electron spectroscopy (AES) is an electron spectroscopy which probes Auger electrons emitted from atoms whose inner shells have been ionized by energetic electron impacts. The energies of the measured Auger electrons bear information about the atomic energy levels of the element from which they have been emitted and about its bonding configuration (through slight changes in the atomic levels associated with bonding). When AES is combined with sputter removal of thin layers of the solid target, this technique offers a sensitive tool for monitoring the depth distribution of the constituent and impurity atoms. A depth scale is established by converting the sputter time to a sputter depth by either measuring the crater formed during the sputtering process or by relying on suitable calibrations. Calibrations must be made on a sample to sample basis as shown by *Elman* et al. because of differences in the impurity species and their concentrations which may affect the sputter yield [5.6]. AES, like other electron spectroscopies, works well for conductive substrates like graphite, but encounters difficulties due to specimen charging when applied to insulators, such as diamond.

Auger electron spectroscopy has been successfully used by various groups as a sensitive tool to monitor the depth distribution of impurities in graphite [5.26, 27, 28]. The use of electron spectroscopies for the characterization of diamond is complicated by its high resistivity. Nevertheless, Auger electron spectroscopy has been applied to diamond for depth profiling as well as for diamond phase identification. The different bonding of the carbon atoms in diamond and graphite is reflected in different Auger electron line shapes [5.29], as deduced from band structure calculations and verified experimentally as shown in Figs. 5.4 and 5.5. A major problem, however, encountered in AES [5.30] when combined with sputtering for depth profiling and identification of the diamond phase is the tendency of the bombarded diamond surface to transform to graphite due to the damage that the sputtering process inflicts on the first few atomic layers of the bombarded diamond surface. This, when combined with the extreme surface sensitivity of the Auger electrons to the environment of the

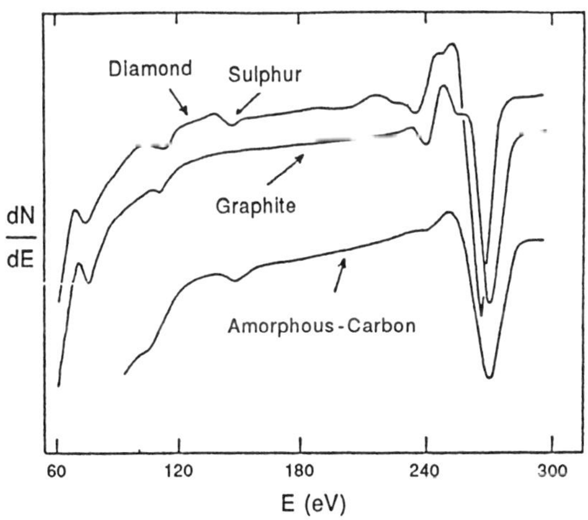

Fig. 5.4. Auger spectra of diamond, graphite and amorphous carbon [5.26]

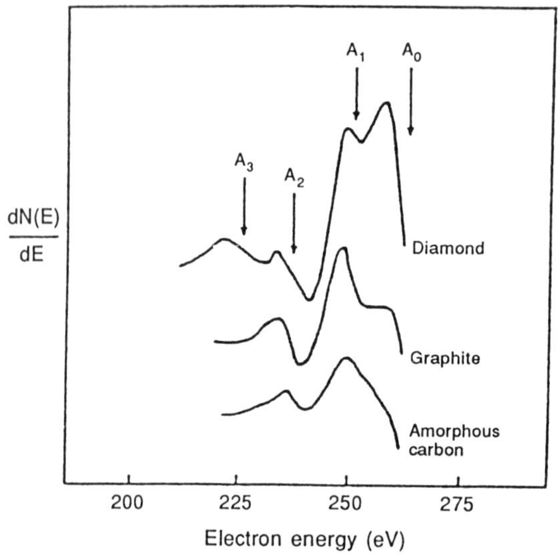

Fig. 5.5. Fine structure of the carbon Auger spectra from diamond, graphite, and amorphous carbon. The main Auger features for diamond are indicated [5.26]

emitting atom, makes a sputter-cleaned diamond surface often look graphitic, as deduced from its AES lineshape.

Very recently, *Hoffman* et al. [5.31] have shown that it is possible to use secondary electron emission spectroscopy (SEE) in the 0 to 60 eV range to provide signatures for different allotropic forms of carbon. Because of the very low energy of the emitted electrons, this technique is confined to the characterization of the first few atomic layers from the surface. In general, the SEE spectrum consists of a broad peak centered at a few eV from the vacuum level, and upon this peak is superimposed a fine structure which reflects the unoccupied density of states above the vacuum level. The SEE spectra of HOPG [5.32] and diamond [5.33] observed by *Hoffman* and coworkers yield good agreement with other spectroscopies sensitive to the unoccupied density of states and with band structure calculations. The SEE spectrum has also been measured for glassy carbon, sp^2 bonded amorphous carbon (a–C), and amorphous diamond (i.e., sp^3 bonded amorphous carbon) [5.34, 35]. The SEE signatures for these allotropes are easily distinguishable from each other, more so than the corresponding C(KLL) Auger lineshapes. A useful comparison of the sensitivity of AES, Raman spectroscopy and SEE is provided by measurements applying each of these three techniques to the surfaces of CVD deposited diamond films [5.36]. SEE, being a surface-sensitive technique, was found to be consistent with, but more sensitive than, the Raman and Auger techniques for the assessment of the degree of crystal perfection of the diamond surface.

Electron energy loss spectroscopy (EELS) has also been used by *Prawer* et al. to obtain information on the electronic structure within the first 20 or 30Å from the surface, and subtle changes in the spectra have been used to

Fig. 5.6. Atomic resolution image of the carbon atoms in graphite obtained with an atomic force microscope [5.43]

obtain information on the relative amounts of sp^2 and sp^3 bonding close to the surface [5.25].

Scanning tunneling microscopy (STM) provides an electron microscopy technique with atomic resolution suitable for the study of both the surface topography and local density of states at the Fermi level of a solid surface [5.37]. By using the constant current mode in the STM, the surface topography can be investigated and individual defects or defect clusters can be imaged. The geometrical excursions of the probe tip above the surface while scanning in the constant current mode define the corrugation amplitude and yield detailed information on the reconstruction of the surface atoms. Unlike conventional electron microscopy instruments (such as TEM, SEM, STEM), the scanning tunneling microscope (STM) also provides information on the electronic structure using the spectroscopic dI/dV mode, whereby the local density of states can be scanned and correlated with the topographic features. Because the tunneling tip interacts with the the tails of the electron wave functions which extend out of the surface, detailed theoretical calculations are needed for quantitative interpretation of the observed patterns and their identification with specific defect structures [5.38-40]. Whereas the scanning tunneling microscope is used for conducting samples, the closely related atomic-force microscope, which is sensitive to the force between individual atoms, can be applied also to insulating samples.

Since graphite cleaves so easily, extended flat surfaces can be prepared by cleaving single crystals or HOPG. Thus graphite has become a prototype material for STM studies (Fig. 5.6). As a result, crystalline graphite and HOPG surfaces have been extensively characterized by many groups using this technique [5.40-42]. Some direct measurements of the forces between the tip and a graphite surface have also been made during the STM image scans [5.43]. Some STM studies on carbon fibers [5.35, 42, 44-46] and graphite intercalation compounds [5.47-49] have also been undertaken by a number of groups. Recently, scanning tunneling microscopy (STM) has been used by *Coratger* et al. to gain information on the surface damage created by ion implantation into graphite [5.50].

The corrugation amplitude of a clean graphite surface is about 1 Å as the tip moves from the unoccupied center of the graphene honeycomb structure to an A or B site. Because of the small difference in potential energy of the carbon atoms on A and B sites, there is a small elevation of the B sites (by about 0.1 Å) relative to the A sites at the surface. Therefore, an STM image of the topography of a graphite surface focuses on the B sites [5.51-54] , and the resulting STM image of a perfect graphite lattice appears as a triangular net. For an isolated graphene layer (which for example occurs in an incommensurate graphite intercalation compound where sequential graphene layers are uncorrelated), the A and B sites become equivalent and the STM image of the graphene layer becomes a honeycomb structure with 6-fold symmetry, as shown by the work of *Olk* et al. [5.55]. Although little work has thus far been done with the STM to study ion implantation-induced defect structures, and no work has yet been reported on the use of this technique for regrowth studies, the technique offers great promise for future work.

5.4 X-Ray-Related Characterization Techniques

Various X-ray techniques can be or have been used for the characterization of ion-implanted materials in general and carbon-based materials in particular. For example X-ray diffraction can yield information on implantation-induced changes in lattice constants and in crystallite sizes in highly damaged materials by observing the positions of the X-ray diffraction maxima and their linewidths, respectively. In more heavily damaged materials, for which the crystal structure in the implantation-affected volume may be essentially lost, measurements of the radial distribution function may yield valuable information on the implantation-induced changes in the nearest-neighbor and higher-order distances [5.25]. It should however be noted that an inherent difficulty with X-ray diffraction techniques for many radiation damage assessment measurements is the relatively large probing depth of the X-rays, which often greatly exceeds the implantation-affected depths.

Apart from the X-ray diffraction techniques which are used to gain structural information on the degree of crystallinity of the material under study, there are several other techniques which use X-rays in a completely different way to obtain information about the elemental composition of the specimen. These all rely on the detection of characteristic X-rays emitted from constituent atoms of the target. One of these techniques, namely PIXE, has already been mentioned in Sect. 4.3. Other techniques that are closely related to PIXE and only differ in the mode by which the atomic excitation is achieved are the X-ray fluorescence and the energy dispersive X-ray analysis (EDX) methods. In a fluorescence experiment the excitation of atoms in the target is achieved by irradiating the specimen with photons having energies which exceed the energy required to ionize a specific atomic shell of the element under study. The photons (characteristic X-rays) which are emitted while the holes in this shell are filled up by some outer electrons are detected just like ions in a PIXE experiment in which the X-rays generated by ions are measured. A very similar process is that used in EDX; however, here the atomic ionization is achieved by impact with energetic electrons, usually obtained in an electron microscope.

All three techniques, PIXE, fluorescence and EDX, have very high elemental resolution (determined by the resolution of the detector used for the X-ray counting), and all suffer from inferior depth resolution due to the gradual decrease in X-ray production and escape probability with depth in the target. PIXE has the advantage that, when combined with the ion-channeling effect, it can also yield structural information (location on the lattice of the impurity atoms). EDX, when combined with the electron scanning feature of a scanning electron microscope (SEM), can give information on lateral mass distributions at the high resolution that a SEM can offer, while X-ray fluorescent experiments are easy, and readily available. All these techniques when applied to carbon-based materials exhibit high sensitivities to the detection of impurities in the carbon matrix, due to the particularly low atomic number of carbon which gives rise to characteristic X-rays of extremely low energy. Thus it is easy to eliminate the characteristic carbon X-rays from the recorded X-ray spectrum.

5.5 Electronic Transport Measurements

In some cases electronic transport measurements can be used to characterize ion-implanted materials. In the case of diamond and a–C:H (diamond-like) films, where the unimplanted material is highly resistive and the implantation process turns it electrically conductive, transport measurements provide a sensitive probe of the near-surface modifications of the material. On the other hand, for graphite and graphite-related materials, ion implantation causes an increase in the electrical resistivity of the near-surface region, which is difficult to measure in the presence of a relatively more conducting substrate. By working with very thin specimens and by devising special electrodes, the implantation induced changes in the surface resistance of graphite can be sensitively measured (Sect. 6.4). In the case of graphite, ion implantation to very high doses eventually forms a material with a resistivity similar to that of amorphous carbon ($\rho \sim 10^{-2}$ Ω cm). In addition to conductivity measurements, other transport techniques, such as Hall effect, magnetoresistance, and thermopower measurements, have been successfully used to characterize ion-implanted graphite, diamond and diamond-like films. By performing transport measurements as a function of temperature, important information on the electronic conduction mechanism can be deduced.

As will be shown in detail in Sect. 8.1, ion-implantation of diamond may lead to several competing processes which all may give rise to dramatic increases in electrical conductivity. These are (i) graphitization which, at high enough doses, will lead to the formation of a connected conduction pathway, (ii) agglomeration of the implants which, if conductive by themselves, may at high enough doses give rise to conductivity due to hopping between precipitates of the implanted species, and (iii) doping which can in principle give rise to p or n-type conductivity. A variety of experimental techniques have been used to discriminate between these conduction mechanisms – the most fruitful of them being measurement of the temperature, dose and ion species dependence of the conductivity, as will be described in Sect. 8.1.

5.6 Electron Spin Resonance

Electron spin resonance (ESR) provides a key tool for examining the effective g-factors for both conduction and localized (defect) electron states in materials in which not all electrons are paired. In general, ESR measures electronic g-factors through the Zeeman interaction

$$\hbar\omega = g\mu_B H_0 \langle S \rangle \tag{5.1}$$

as the electrons are excited from the ground state in a spin flip transition. The experiments are normally carried out in a microwave resonant cavity at constant frequency $\hbar\omega$ with the magnetic field being swept through the resonant field H_0. Typically, the derivative of the ESR resonance curve is plotted. From the magnitude of H_0, the g-factor in the direction of $\mathbf{H}_0$ is determined using (5.1), where μ_B is the Bohr magnetron and $\langle S \rangle$ is the spin value. Furthermore, from the magnitude of the ESR signal, information about the number of unpaired spins in the specimen can be deduced. ESR measurements are therefore particularly sensitive to the assessment of implantation-induced defects (broken bonds) and to following their annealing. Such measurements have indeed been extensively applied to graphite, diamond and related materials.

Since ESR experiments are carried out at microwave frequencies, the penetration depth δ of the electromagnetic fields into the sample depends on the electrical conductivity σ through the relation

$$\delta = \frac{c}{[2\pi\mu\sigma\omega]^{1/2}} \tag{5.2}$$

and δ (through its dependence on σ) may vary as the material is modified by ion implantation. Prior to implantation, the skin depth δ for diamond for example is large compared to the sample size but, as the ion dose increases, σ increases and δ approaches the value of the ion penetration depth. On the other hand, ion implantation in graphite typically reduces σ, so that δ increases. Since $\delta \gg R_p$ in many cases (especially in diamond), there are often contributions to the ESR spectrum from the unimplanted regions of the sample well beyond R_p which are still within the skin depth δ. Furthermore, because of the inhomogeneities in composition and defect density in ion-implanted diamond, graphite and related

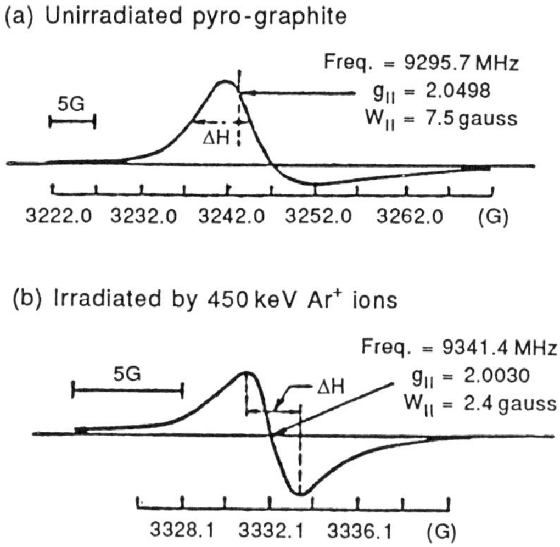

Fig. 5.7. (**a**) A typical ESR derivative lineshape for a well-graphitized sample. The actual trace is for a pyro-graphite sample prior to ion implantation. From the frequency of the cavity (9295.7 MHz) and the central ESR frequency, the g value for $H \parallel c$-axis is found (2.0498 for the sample shown). The width of the ESR line in the figure is $\Delta H = 7.5$ gauss. (**b**) A typical ESR derivative lineshape for a highly disordered graphite sample with a high density of localized spins. The actual trace is for the same pyro-graphite sample as in (**a**) after irradiation by 450 keV Ar^+ ions. After irradiation the g-value approaches the free electron value and the linewidth decreases substantially [5.61]

carbon materials, the magnetic field distribution in an ion-implanted sample may be non-uniform, giving rise to a broadening of the ESR line.

Conduction electrons, holes and localized carriers, all contribute to the ESR spectrum [5.56-59]. For the case of graphite, the ESR lineshape is dominated by charged carriers, while that for disordered graphite, diamond and related materials, all exhibit contributions also from localized spins associated with dangling bonds at the defect sites. Clearly, perfect diamond, with all tetrahedral bonds satisfied, does not show any ESR absorption. For disordered graphite the carriers and localized spin contributions appear either as individually-resolved ESR lines or as a superposition of unresolved contributions to a single ESR line, for which a lineshape analysis must be carried out to identify the two contributions separately [5.60]. Such an analysis is possible because the lineshapes for the two contributions are different, arising from the different mechanisms for the two cases. Figure 5.7a illustrates the ESR lineshape in graphite, and its asymmetric (Dysonian) [5.62] lineshape, where the asymmetry arises from the carrier conductivity within the skin depth and has been characterized by *Wagoner* in terms of the ratio of B/A, the ratio of the positive peak intensity to the negative peak intensity [5.56]; the B/A ratio approaches unity as $\sigma \to 0$. Thus as disorder is introduced into graphite by ion implantation, σ decreases and the B/A ratio is reduced. Disorder (as might be caused by ion implantation) gives rise to localized states, which yield a symmetric Lorentzian ESR lineshape, characteristic of an insulator (Fig. 5.7 where the derivative is plotted). A similar lineshape has been observed by *Adel* et al. for as-grown and ion-implanted a–C:H [5.63].

The ESR lineshape and width, even when symmetric, may bear information about the nature of the interaction between the spins. A Lorentzian lineshape is indicative of exchange (or motional) interactions which the spins undergo, in contrast to a Gaussian lineshape which is caused by dipolar interactions. The ESR lineshape for disordered diamond is found to be Lorentzian and symmetrical. Low levels of lattice damage in diamond introduce dangling bonds and unpaired spins, so that the intensity of the ESR line provides a measure of the concentration of these unpaired spins. Higher levels of lattice damage introduce a sufficient density of charge carriers, thereby resulting in some Dysonian character to the ESR line. The analysis of the ESR lineshape and shifts of the peak positions thus provide information on the amount of disorder in the samples. Another consideration for the interpretation of the ESR lineshape of ion-implanted diamond is the formation of graphitic material through the ion-implantation process, so that the ESR spectrum for ion-implanted diamond is expected to have contributions from undamaged diamond regions, lattice-damaged diamond regions, and defective graphitic material. Specific defect centers give rise to characteristic spectra, and the observation of such spectra can be used to identify the presence and concentration of such defects.

Two parameters characterize the Lorentzian ESR lineshape of the isolated spins, namely the magnetic field H_0 at maximum resonant absorption intensity and the half width at half maximum intensity ΔH. These same parameters are also used to characterize non-Lorentzian lines, where the linewidth ΔH is measured as the difference in magnetic field between the maximum and minimum in the lineshape (Fig. 5.7). From the intensity of the ESR line, the magnetic susceptibility χ can be deduced [5.64] and a fit of the temperature dependence $\chi(T)$ to a Curie law is often used to yield the unpaired spin concentration.

The g-factor (5.1) for an isolated electron spin is 2.0023; therefore departures from this value provide information on the interaction with other spins and conduction electrons. The temperature dependence of the linewidth is especially sensitive to the spin-spin and spin-lattice interactions. The asymmetry of the lineshape provides additional information on the interaction of the spins with the conduction electrons (through the B/A ratio, shown in Fig. 5.7a).

The shift of the g-factor from the free electron value is very large in graphite because of the very high in-plane diamagnetism in graphite due to orbital band structure effects [5.60]. Since lattice disorder decreases the diamagnetism dramatically, measurements of H_0 are very sensitive to the in-plane crystallite size. Furthermore, the g-factor for graphite is highly anisotropic varying from 2.0026 to 2.0495 with a very large diamagnetic shift observed for $g_{\|}$ (where $H \parallel c$-axis) and almost no shift for $g_{\perp}$ (where $H \perp c$-axis); the angular dependence of the g-factor in graphite has been explicitly given by *Wagoner* as $g(\theta) = g_{\perp}+(g_{\|}-g_{\perp})\cos^2\theta$ where $(g_{\|}-g_{\perp}) = +0.125$ at 77 K and $+0.047$ at 300 K [5.56]. As the graphite becomes more disordered, this anisotropy is dramatically reduced. For this reason the anisotropy of the g-factor is another important characterization parameter for lattice damage in ion-implanted graphite and other carbon-based materials. The anisotropy in the g-factor is also sensitive to the extent of the preferred spin alignment.

5.7 Hyperfine Interactions

The interaction of nuclear electromagnetic moments with extra-nuclear fields falls into the general category of hyperfine-interactions (HFI). The two most significant interactions among those are (i) the magnetic interaction of a nuclear dipole moment with hyperfine magnetic fields induced either by atomic vacancies or by the magnetism of the solid environment, and (ii) the quadrupole interaction of a nuclear quadrupole moment (deviation from spherical symmetry of the nuclear state) with electric field gradients induced by asymmetries in the environment; such as caused by a defect in the surroundings of the probe nucleus. As a result of these interactions, a splitting in the nuclear magnetic sublevels (m_I) occurs which can be experimentally detected. Once this hyperfine splitting is measured, the product of the nuclear moment and the environmental field can be deduced. For those cases in which the properties of the nuclear probe are known, information can be extracted about the environment, i.e., in the case of quadrupole interactions, about the presence of damage or about the symmetry of the location of the probe nucleus in the solid matrix.

The two HFI techniques most widely used are the Mössbauer effect and the perturbed angular correlation technique. Both have been applied to the study of defect structures and the annealing of implant-induced damage in carbon-based materials.

5.7.1 Mössbauer Spectroscopy

The basics of a Mössbauer experiment are that a γ-ray is emitted by an excited nucleus of suitable properties located in a solid matrix (the source) and this γ-ray is then resonantly absorbed by the identical nucleus in its ground state, located in a different environment (the absorber). The γ-ray energy is modulated in a Mössbauer experiment by a small varying amount $\pm\delta E$ through Doppler-shifting the source relative to the absorber. This modulation may bring the energy of the γ radiation into a resonance condition, even for those cases in which it does not exactly match the energy level separation in the absorber. These energies in the absorber may differ from those of the source, despite the fact that they belong to the same ground and excited levels in the Mössbauer nucleus, because of hyperfine splittings caused by the environment in which the source or absorber nuclei reside. Of relevance to implantation effects, as deduced from Mössbauer measurements, is the quadrupole hyperfine interaction which gives rise to a quadrupole splitting ΔE_Q originating from the interaction between the nuclear quadrupole moment and the electric field gradient at the nucleus. The absence of any quadrupole splitting indicates that the probe nuclei both in the source and in the absorber reside on "unperturbed" (i.e., spherically symmetric) sites. These are usually associated, for the case of diamond, with substitutions in the perfect (defect free) lattice. The "recoil free fraction", a quantity which can be deduced from the strength of the Mössbauer signal, is related to the tightness of the binding of the probe nucleus to its environment. The most commonly used Mössbauer probe nuclei are ^{57}Fe, ^{133}Cs, ^{125}Te and ^{129}I, all of which have been used to study implantation effects in diamond, and some also in graphite.

5.7.2 Perturbed Angular Correlations

In perturbed $\gamma - \gamma$ angular correlation experiments use is made of the fact that some favorable nuclear decays (like that existing in ^{111}Cd) are characterized by a $\gamma - \gamma$ cascade which proceeds through an isomeric state having non-zero electromagnetic moments and an anisotropic angular correlation. The hyperfine interaction (HFI) which this state may undergo during its lifetime is reflected in a measurable modification of the time evolution of the $\gamma - \gamma$ angular correlation function, $W_{\gamma_1\gamma_2}(\theta, t)$ given by

$$W_{\gamma_1\gamma_2}(\theta, t) = 1 + \sum_k A_k G_k(t) P_k(\cos\theta), \tag{5.3}$$

where γ_1 and γ_2 refer to the two γ rays which make an angle of θ with respect to each other. Here A_k are known parameters which depend on the nuclear spins involved in the measured $\gamma - \gamma$ cascade, $P_k(\cos\theta)$ are the Legendre polynomials and the time dependent $G_k(t)$ are determined by the type of hyperfine interaction between the probe nucleus and its environment. For the case of the electric quadrupole interaction (i.e., the interaction of the nuclear quadrupole moment with an electric field gradient V_{zz}) the angular correlation is periodically modulated with a characteristic frequency which can be extracted from experiment.

The important factors deduced from PAC measurements which are relevant to the present discussion are a set of coefficients f_n which reflect the relative contributions of different interaction strengths (i.e., different electric field gradients) caused by different environments of the probe nuclei. If only f_0 exists (i.e., the angular correlation is "unperturbed"), all probe nuclei are located on sites at which they experience no electric field gradient. For crystals having cubic symmetry (e.g., diamond) and for large impurity atoms, these sites are most likely substitutional sites. Effective annealing of the implantation damage is thus reflected in a high fraction of probe nuclei residing on substitutional "unperturbed" sites. These locations in the crystal should in most cases of impurity doping be correlated with a high degree of electrical activation of the dopants.

The coefficients f_n for $n > 0$ reflect the relative abundance of probe nuclei located on non-substitutional sites in the matrix, but on sites where they experience well-defined perturbations which sometimes can be associated with specific environments in the lattice. By performing perturbed angular correlation (PAC) experiments on single crystals oriented at well-known angles with respect to the gamma detectors, one can extract from the experimental data the exact orientation of the electric field gradients experienced by the probe nuclei in the crystal, and hence, the exact defect configuration in the vicinity of the probe nucleus.

A slight variation of the perturbed angular correlation (PAC) technique is the perturbed angular distribution (PAD) method in which the nuclear state under study is populated via a nuclear reaction rather than by a preceding γ-transition; however, in principle, both the PAC and PAD methods are identical in the information they can yield.

The nuclear probes mainly used for PAC studies in graphite and diamond are ^{111}In and ^{181}Hf, and for PAD experiments ^{19}F is used.

5.8 Mechanical Properties

The unique mechanical properties, i.e., the extreme hardness of diamond and the ultra-high bulk modulus of graphite and other carbon-based materials, have attracted much attention. Attempts have also been made to improve these properties further by ion-implantation techniques (Sects. 6.5, 7.1 and 8.9).

The major properties for which irradiation effects have been evaluated are: (i) frictional and scratch resistance, (ii) toughness and wear (iii) machining performance, and (iv) hardness. A brief outline is given below of the experimental arrangements employed to test these properties.

A typical friction test apparatus [5.65] consists of a loading arm which is caused to move slightly against a strain-gauge force measuring arrangement, when the sample undergoing friction testing is driven slowly beneath a fixed load. Toughness and wear of sharp edges (such as a diamond stylus) can be measured by bringing the specimen into contact, under a well-known load, with a rotating disk made of superhard material such as a stainless steel disk coated with a well-polished TiC thick film. The quantity measured is the time to failure of the specimen for a series of different loads, and this time-to-failure has been shown by *Hartley* et al. to provide a measure of the wear resistance of the sharp edge [5.66]. A similar test can be performed under industrial conditions, to evaluate the performance of diamond-tipped high-speed cutting tools with or without ion-implantation treatment. The static hardness of materials is measured by indentation tests. In these tests a sharp edge of well-known material (usually made of diamond) is pressed under a well-known load against a flat polished surface of the material whose hardness is to be evaluated. The indentation depth caused by this test is measured under a microscope and is related to the hardness of the material using well-established procedures (i.e., Knoop hardness). Needless to say, microscopic observations using either optical or electron microscopy are frequently employed for qualitative evaluation of the damage inflicted to or by the material under study.

6. Implantation-Induced Modifications to Graphite

In this and the following chapter, a review of implantation-induced modifications of the structure and properties of graphite (Sect. 6.1–Sect. 6.8) and graphite-related materials (Chap. 7) is presented. Chapters 8 and 9 present a similar review of ion-implantation studies in diamond and diamond-related materials. Due to the variety of bonding characteristics, crystalline structures and thermodynamic properties, the family of carbon-based materials exhibits a rich variety of implantation-related phenomena. The effects of implantation-induced damage, the subsequent annealing of this damage in these materials, and the incorporation of the implanted species in the host lattice are all discussed in the following sections. Particular attention is given to the disparities and commonalities in the behavior of these materials upon implantation, demonstrating the uniqueness of carbon as a host material for ion implantation.

6.1 Lattice Damage

Graphite is an interesting prototype material for ion implantation studies because of the anisotropy in its physical and structural properties. In particular, due to the layered nature of the graphite lattice, the processes of amorphization and recrystallization (graphitization) of the implanted region are conceptually different from that observed in commonly studied materials [6.1]. Since ion implantation provides a controlled method for the introduction of lattice defects which can be subsequently annealed, this technique provides fundamental information on the unique crystalline regrowth (graphitization) process of this highly anisotropic material. Despite the large structural anisotropy of graphite, models developed to describe the implant profile in cubic semiconductors have been shown by *Elman* et al. to be applicable to graphite, based on the results of depth profiling measurements (Sect. 5.3) in highly oriented pyrolytic graphite (HOPG) implanted with a variety of ion species [6.2].

The amount and depth distribution of the implantation-induced defects in bulk graphite (e.g., HOPG) can be sensitively characterized using the RBS-channeling technique. TEM studies on ion-implanted HOPG provide information on the microstructure in the basal plane of the graphite lattice as well as on extended defects with large spatial extent such as dislocations. Complementary information is provided by TEM studies on vapor-grown graphite fibers that have been annealed to high temperatures ($T_{HT} > 2800°C$) to achieve a high

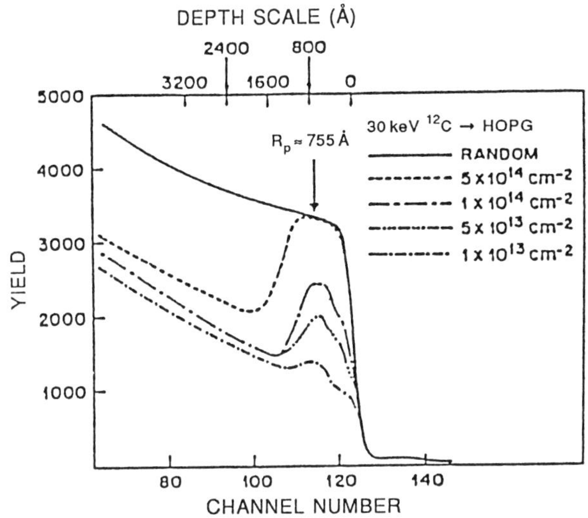

Fig. 6.1. Backscattering spectra along the c-axis channeling direction obtained by *Venkatesan* et al. with 2.0 MeV $^4He^+$ ions on HOPG, implanted with ^{12}C at E = 30 keV to various fluences. A spectrum taken along a random direction is shown for comparison [6.3]

degree of structural perfection. By exploiting the special geometry of the vapor-grown graphite fibers, it is possible to use the high resolution TEM to image the c-axis lattice planes, thereby permitting characterization of the microstructure along this axis. In addition, both HOPG and vapor-grown carbon fibers are amenable to study by Raman scattering, ESR, SEM and a variety of other characterization tools, thereby providing additional complementary information on the lattice defects introduced by ion implantation into graphite.

The creation of lattice damage by the implanted ions can be sensitively and conveniently monitored using RBS-channeling, which provides an overview of the damage-generation process in ion-implanted graphite. By selecting carbon atoms for the implantation projectiles, only lattice damage and no impurity effects are introduced. Figure 6.1 shows the RBS channeling spectra for 30 keV C^+ ions implanted into HOPG to various fluences, using 2.0 MeV $^4He^+$ ions as the RBS probe [6.3]. With this sensitive characterization technique, some lattice damage can already be seen at a fluence as low as 1 $\times 10^{13}$ ions/cm^2 (Fig. 6.1), the initiation of the lattice damage occurring near R_p, the peak in the profile for the implanted species. With increasing fluence, the RBS channeling yield increases to approach that for a random direction, indicative of an increase in the number of blocked channels. The RBS yield (Fig. 6.1) also broadens for depths both greater than and less than R_p. Finally, at a dose of $\sim 5 \times 10^{14}$ C ions/cm^2, many channels extending from the surface to a depth of $\sim R_p + \Delta R_p$ are blocked, as shown by the overlap between the RBS yield for the implanted sample in the channeling direction (the c-axis) and in a random direction (Fig. 6.1) [6.3].

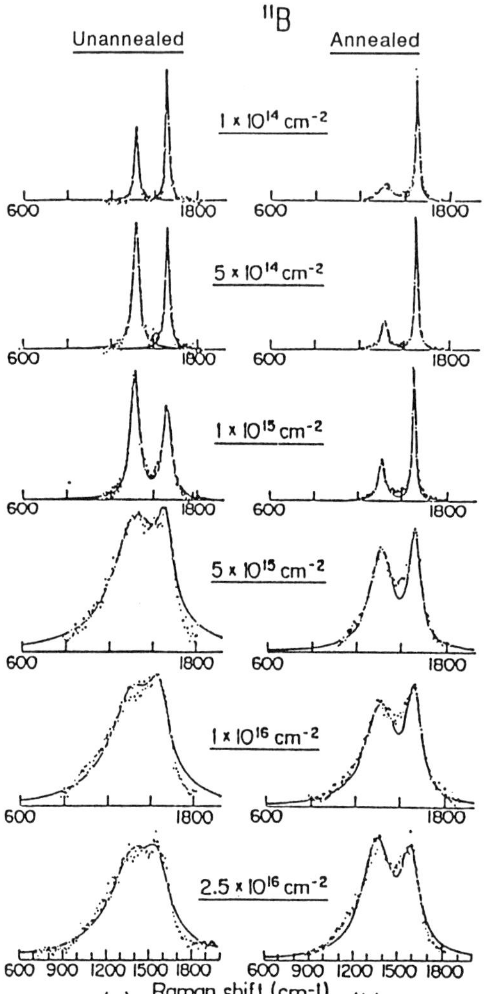

Fig. 6.2. Raman spectra measured by *Elman* et al. for various fluences (1×10^{14} to 2.5×10^{16} ions/cm^2) of ^{11}B ions implanted into six HOPG samples. The spectra on the left (**a**) are for the unannealed samples, and on the right (**b**) are for the same samples annealed at 950°C for 0.5 h [6.4]

Similar conclusions about the effect of ion dose on lattice damage are reached from Raman scattering studies. Figure 6.2 shows the effect of ion dose on the Raman spectra for graphite implanted with ^{11}B ions at 100 keV. The Raman measurements are sensitive to the lattice damage present within the optical skin depth of the probing laser radiation (Sect. 5.1), particularly to the near-surface region through which the energetic ions and their knock-on ions pass before coming to rest. For relatively low fluences (e.g., $\phi \sim 1 \times 10^{13}$ cm^{-2} in Fig. 6.1) where the lattice damage is mostly confined to the region where the ions stop (between $R_p - \Delta R_p$ and $R_p + \Delta R_p$), well-resolved Raman lines are observed for both the G and D lines of the Raman spectrum and average values for L_a within the optical skin depth have been determined by *Elman* et al. from the ratio of the intensity of the D and G lines (Sect. 5.1) [6.4].

As the ion dose increases and the region of heavy lattice damage moves towards the surface (Fig. 6.1), both the G and D Raman lines broaden rapidly with increasing dose, so that L_a can no longer be sensitively determined from the measured line intensities of the Raman spectra (Fig. 6.2). For example, the in-plane crystalline-size L_a decreases from $L_a \sim 1\mu$m for the unimplanted material to $L_a \sim 75$Å for a dose of $10^{14}/\text{cm}^2$ $^{11}B^+$ ions incident at 100 keV on HOPG, and to $L_a \sim 30$Å at a dose of $10^{15}/\text{cm}^2$ [6.4]. For higher doses, the damage is too great to allow the extraction of a reliable estimate of L_a from analysis of the Raman lineshape [6.4, 2]. For example, for a dose of 5 $\times 10^{15}/\text{cm}^2$ $^{11}B^+$ ions incident at 100 keV, the Raman lineshape is similar to that of amorphous carbon (compare Fig. 6.2 with Fig. 5.1) [6.4]. A theoretical calculation by *Lespade* et al. of the Raman spectrum expected for disordered graphite [6.5] shows that the linewidths characteristically increase with disorder, since contributions from a larger range of wave vectors and bond angles must be considered [6.5, 6]. Shifts in the peak frequencies are also observed upon ion implantation [6.4] since the phonon frequencies corresponding to this wider range of wave vectors and bonding angles are not symmetrically located with respect to the zone center Raman-allowed phonons or to the maximum in the phonon density of states (Fig. 2.10).

Although resonant Raman effects [6.7-12] have often been used to emphasize the D line (Sect. 5.1) relative to the G line in disordered graphite-based materials, this effect has not yet been applied to study ion-implanted graphite. The published resonant Raman studies show an enhancement of the ratio I_D/I_G as the laser frequency decreases. This effect is associated with enhanced interband intensity in the lower photon energy region for disordered carbons, thereby "resonantly" enhancing the contributions from the more disordered regions of the sample relative to the more ordered regions at lower laser excitation energies E_L.

Of all the techniques that have been used to study defect structures in ion-implanted graphite, scanning tunneling microscopy (STM) provides information with the greatest spatial resolution [6.13] (Sect. 5.3). However, since STM is limited to probing the surface layer and the next nearest layer only, this technique is not well matched to the implant and damage profiles associated with typical ion-implantation experiments. The most common defect structures that are observed in STM correspond to the intersection of implantation-induced cascades with the surface. One of the early STM studies on ion-implanted graphite was carried out by *Coratger* et al. on freshly cleaved HOPG, implanted at room temperature with 20 keV $^{12}C^+$ ions to fluences in the range 1×10^{12} to 4×10^{15} ions/cm^2 [6.13]. The $^{12}C^+$ implant was chosen to avoid donor or acceptor charge transfer effects which would complicate the interpretation of the STM images.

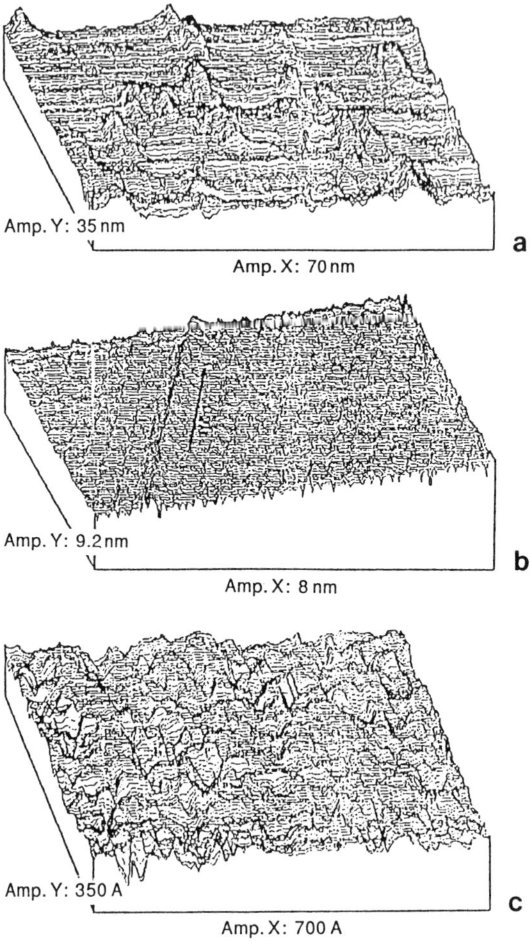

Fig. 6.3. STM images taken by *Coratger* et al. in the constant current mode for ion-implanted graphite. (**a**) Isolated hillocks are seen at low doses (2×10^{12} C ions/cm^2 at 20 keV). (**b**) Ridged surfaces are seen along a [1100] direction for intermediate doses such as $\phi = 1 \times 10^{13}$ cm^{-2}. (**c**) The characteristic graphite surface structure is lost for high ion fluences (4×10^{15} ions/cm^2) [6.13]

Under these implantation conditions, values for the implantation parameters are $R_p = 470$Å, $\Delta R_p = 110$ Å and the peak of the damage zone is 300Å below the surface. For a dose of 1×10^{14} ions/cm^2, the maximum concentration of the implant corresponds to 4×10^{19} ions/cm^3 with the implant density being about 4 orders of magnitude lower at the surface [6.13].

The STM images of ion-implanted graphite show dramatic implantation-induced changes in the topographic features which are strongly dependent on the ion dose, as shown in Fig. 6.3. For low doses ($\approx 1 \times 10^{13}$ ions/cm^2), localized positive corrugations about 5 times the magnitude (≈1Å) of that observed for HOPG prior to implantation are found (Fig. 6.3a) [6.13]. This enhanced corrugation is attributed to lattice rearrangement or relaxation following defect migration associated with an isolated cascade. To observe the effect of individual ion impacts on the sample, very low ion doses ($\sim 10^{11}$ argon ions at 50 keV) have been used, and the scans in this case show a single hillock associated with the lattice strain introduced by a single ion impact on the surface [6.14, 15]. For intermediate doses (1×10^{13} to 1×10^{14} ions/cm^2 C ions at 20 keV), isolated corrugation lines (about 1–4 Å high, about 2Å wide and about 10Å long) are observed along the [1100] direction (the direction of maximum atomic density), as shown in Fig. 6.3b [6.13]. These corrugation lines arise from the formation of an extra row of atoms near the surface. These extra rows are thought to result from atomic rearrangements initiated by subsurface cascades which cause internal stresses that are relieved by volume expansion and atomic displacements along certain preferred directions [6.13, 15]. Surface ridges were also observed on a micron scale by *Elman* et al. [6.2] using SEM techniques, but at much higher doses (exceeding 10^{15}/cm^2 $^{11}B^+$ ions). Higher magnification STM scans show that in the vicinity of the enhanced corrugation features, the surrounding atoms show large perturbations from their equilibrium sites [6.13]. Above a dose of 1×10^{14} ions/cm^2, the surface atoms become highly perturbed (due to the overlapping cascades) and show a mean surface roughness of 30Å. In this regime, the characteristic structure of the graphite surface is lost, as shown in Fig. 6.3c [6.13].

For all dose levels, the density of surface defects in the observed region is less than the density of the ionic impacts at the surface and, furthermore, the density of surface defects does not scale with the dose [6.13]. The observation of protrusions rather than depressions is consistent with the addition of ions and the possible spatial imbalance between vacancies and interstitials created by the implantation process, both of which cause volume expansion. This observation is possibly also related to the creation of separated vacancy and interstitial-rich regions, as described in Sect. 8.2 in connection with the observed volume expansion in implanted diamond. High resolution images of the defects by *Mizes* et al. show that the ion-induced defects differ from those produced by substituted impurities [6.16].

The amorphization of HOPG under the influence of 1 keV Ar^+ ions has been studied in-situ using secondary electron emission spectroscopy (SEE) [6.17] (Sect. 5.3). The gradual transformation of HOPG to an amorphous phase was monitored by observing the gradual washing out of the fine structure of the SEE spectrum [6.17, 18]. As estimated by SEE, a dose of 4×10^{14} Ar^+/cm^2 at 1 keV was required to amorphize the HOPG surface at room temperature.

Complementary information on the nature of the implantation-induced damage is provided by ESR studies which give direct evidence (through the ESR lineshape and the temperature dependent susceptibility, see Sect. 5.6) for localized spins created by the implantation-induced lattice damage. These localized spins are associated with pockets of disorder resulting from cascading clusters [6.19]. The residence time of spins in localized states becomes longer with

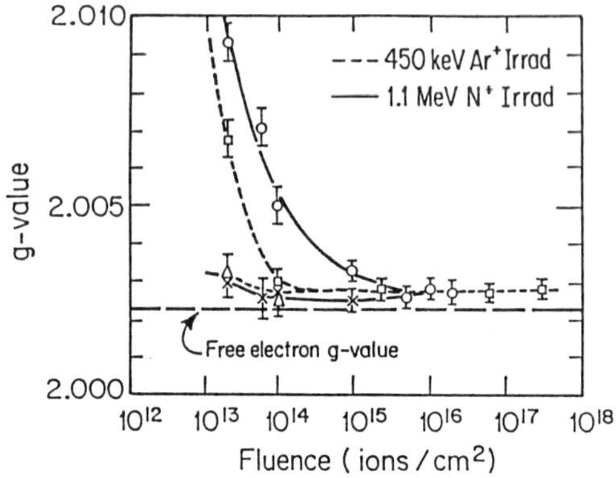

Fig. 6.4. Room temperature g-values versus fluence reported by *Kazumata* et al. for pyro-graphite irradiated by 450 keV Ar^+ ions (*dashed curves*) and by 1.1 MeV N^+ ions (*solid curves*). The large g-values (○ and □) correspond to $g_{||}$. The g-values (△ and ×) close to the free electron value of 2.0023 correspond to $g_{\perp}$ [6.19, 20]

increasing concentrations of defects, giving rise to a shrinkage of the wave functions associated with spins. In this connection, using ESR as a characterization probe of the lattice damage, *Kazumata* and *Yugo* [6.19] showed (Fig. 6.4) that the ESR technique could be employed to quantitatively monitor the transition from the conduction carrier regime (characteristic of graphite) to the localized spin regime (characteristic of amorphous carbon) [6.19]. Increasing the fluence was found to increase the intensity of the ESR line, to reduce $g_{||}$ dramatically, and to decrease the anisotropy in the g-factor; specific results (Fig. 6.4) were obtained for Ar^+ ion implantation at 450 keV (maximum displacements per atom [dpa]=5.6) and N^+ ion implantation at 1.1 MeV (maximum dpa=1.6) [6.20]. It was found by *Kazumata* et al. that, for doses in the $10^{14} - 10^{15}/cm^2$ range, $g_{||}$ approaches the free electron value as the anisotropy in the g-shift disappears, signaling amorphization. At the same values of fluence where saturation in $g_{||}$ occurs, the linewidth for the ESR lines becomes narrowed to its saturation value, as the conduction electron contribution is quenched so that only contributions from the localized spins are observed [6.21].

The ESR spectra for higher energy (> 1 MeV) implants characteristically show a superposition of lines associated with different defect centers [6.21]. Also characteristic of these higher energy implants is a near-surface region with a relatively low density of point defects. This near-surface region can be sensitively probed by the ESR experiment. Point defects can be relatively well annealed as the annealing temperature is increased to $T_a \sim 1000$ K. For these high energy implants, the two ESR characterization parameters, the g-shift and the ESR linewidth, both show a dependence on measuring temperature that is characterized by an activation energy associated with the hopping of spins [6.21].

The process of accumulation of implantation-induced lattice disorder in the graphite structure, as characterized by the RBS, Raman, and ESR measure-

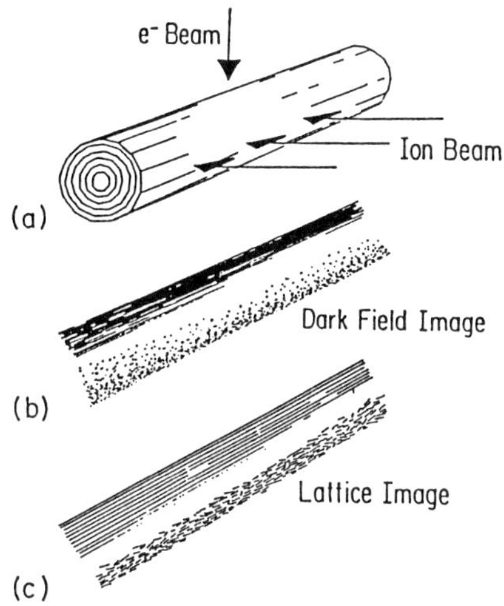

Fig. 6.5. Schematic representations of (**a**) the directions of the ion beam for implantation and the electron beam for TEM observations. Schematic view for (**b**) the (002) dark field and (**c**) lattice fringe images of the implanted and unimplanted sides of the fiber observed in the transmission electron microscope [6.23]

ments, does not differ greatly from that observed in isotropic materials. However, there are some peculiarities in this process which are clearly revealed in TEM studies of ion-implanted vapor-grown (benzene-derived) graphite fibers [6.3, 22, 23] and these differences in the accumulation of induced lattice disorder can be correlated with the anisotropic nature of graphite. Fiber host materials are used to provide submicron sample thicknesses without the need to apply thinning procedures which tend to introduce complicating artifacts into the TEM observations of highly anisotropic materials. By using a suitable low energy of implantation, the implantation-induced damage can be confined to a shallow region near the surface of the fibers, thereby permitting the simultaneous observation (by TEM) of the damaged zone and of the pristine substrate. Figure 6.5a [6.23] shows schematically the directions of the ion beam for implantation and the electron beam for microscopy characterization [6.22, 23], as well as the schematic (002) dark field (Fig. 6.5b) and lattice fringe images (Fig. 6.5c) for both the implanted and unimplanted sides of a graphite fiber.

Figure 6.6a shows dark field TEM images of graphite fibers indicating that the pristine benzene-derived graphite fibers heat treated to $> 2800°$C exhibit large areas of straight and defect free graphite layers extending over 1000Å along both the a-axis and the c-axis directions. Thus benzene-derived fibers provide a good approximation for single crystal graphite for TEM applications. The interlayer spacing is determined to be 3.36Å from optical diffractograms taken from the negatives of the (002) lattice images (Fig. 6.6a) and also from the (002) X-ray diffraction line using CuK_α radiation. The three-dimensional stacking order of the graphite layers in the fibers prior to implantation is determined

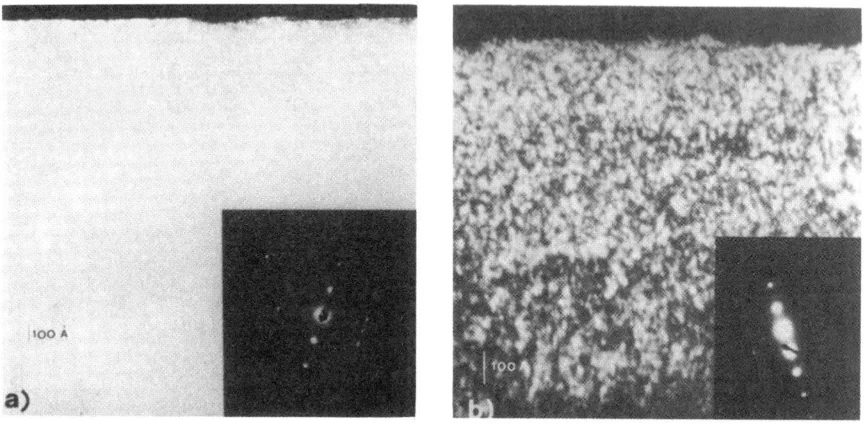

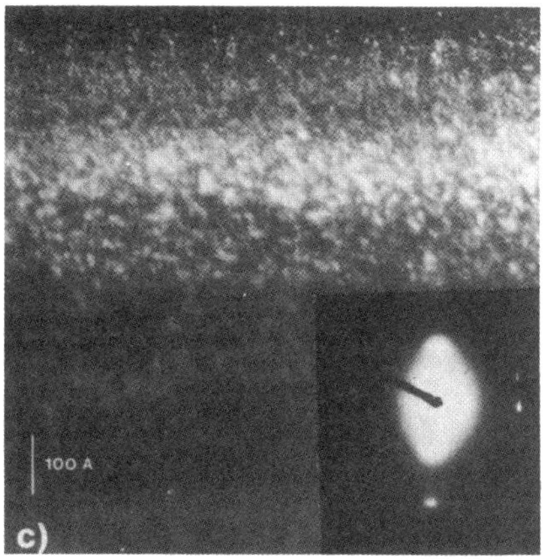

Fig. 6.6. (002) dark field TEM images by *Endo* et al. of (**a**) a pristine fiber. TEM (002) dark field images are shown for fibers implanted with (**b**) ^{75}As and (**c**) ^{209}Bi ions to a dose of 1×10^{15} ions/cm^2 at 30 keV. The insets are the respective electron diffraction patterns [6.23]

from observations of both the (112) diffraction spots of the selected area electron diffraction pattern and the (112) X-ray diffraction line.

The effect of the implant mass on the structural damage of the ion-implanted fiber is shown graphically in Fig. 6.6 by use of the (002) dark field TEM technique [6.23]. The corresponding selected area electron diffraction patterns are shown in the insets. In this figure the results for ^{209}Bi (heavy ion) and ^{75}As (lighter ion) implantation at room temperature at an energy of 30 keV and a fluence of 1×10^{15} ions/cm^2 are shown in comparison with each other and with the (002) dark field image for the highly ordered unimplanted pristine fiber. Ion implantation with ^{209}Bi and ^{75}As under these implantation conditions reduces the crystallite sizes to dimensions as small as 20Å and 50Å, respectively. The projected ion penetration depths were $R_p \sim 147$ Å and $\Delta R_p \sim 26$ Å for the ^{209}Bi, and $R_p \sim 170$ Å and $\Delta R_p \sim 48$ Å for the ^{75}As. Figure 6.6 clearly shows that heavier ions yield smaller crystallite dimensions after implantation, indicating that the heavier the ion, the greater the damage to the graphite structure.

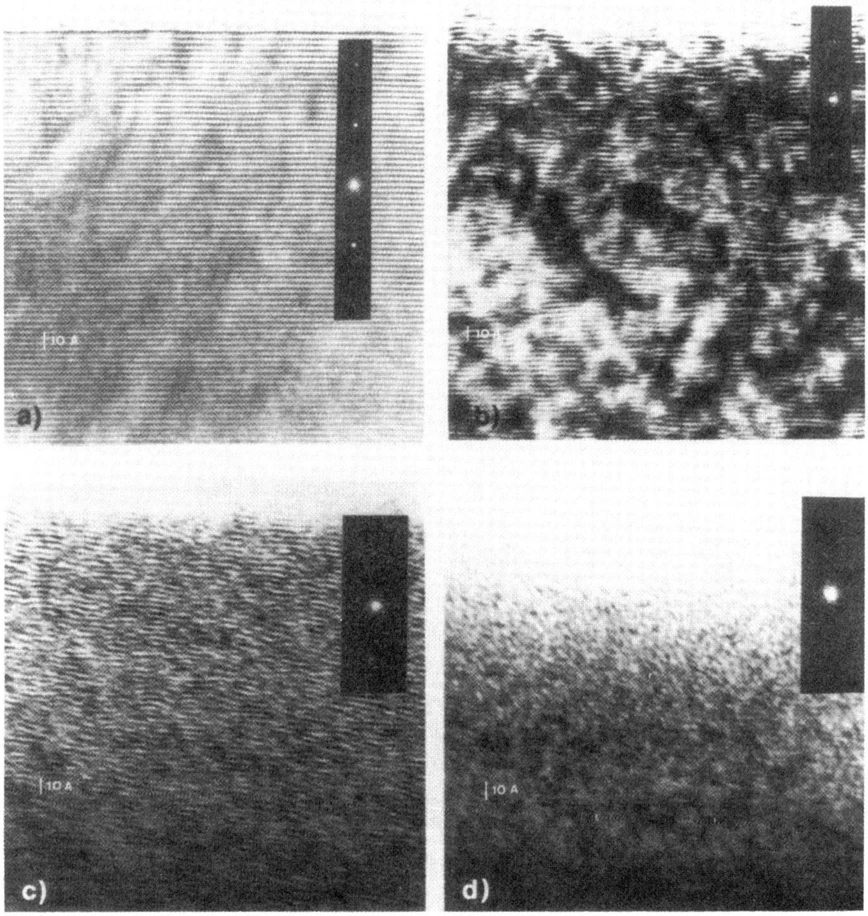

Fig. 6.7. (002) lattice images of (**a**) a pristine vapor-grown fiber and similar fibers implanted with ^{121}Sb ions at 30 keV for various fluences: (**b**) 5×10^{12}, (**c**) 1×10^{14}, and (**d**) 1×10^{15} ions/cm^2. Here, R_p ~151 Å and ΔR_p ~35 Å. The insets are optical diffractograms taken from the negatives of the corresponding lattice images [6.23]

Implantation-induced misalignment of the crystallites with respect to the fiber axis, and an increase of the interlayer spacing from 3.36Å to as much as 3.9Å are deduced from the angular spread of the arced and diffuse (002) electron diffraction spots (not shown) [6.23].

To complement the dark field (reciprocal space) images in Fig. 6.6, real space (002) lattice images of pristine and 30 keV ^{121}Sb ion-implanted fibers are shown in Fig. 6.7 for various fluences from 5×10^{12} to 1×10^{15} ions/cm^2 [6.23]. It is clearly observed in this figure that, with increasing influence, there is a decrease in both the in-plane crystallite size (L_a), as measured from the length of the fringes, and in the thickness of the crystallites (L_c), as measured from the number of parallel stacked layers. At the highest fluence of 1×10^{15} ions/cm^2, the

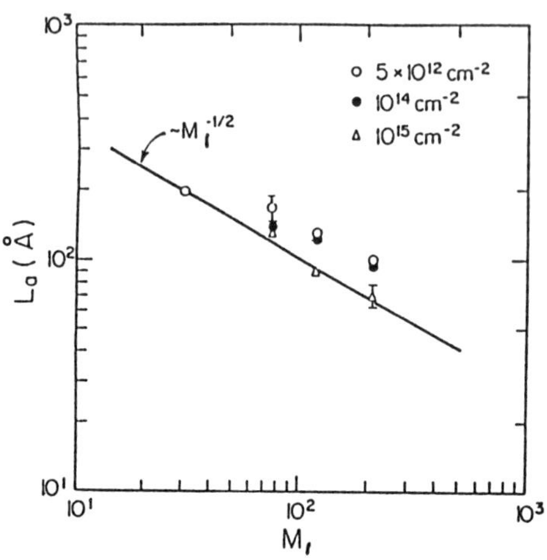

Fig. 6.8. Dependence of in-plane crystallite size L_a on implant ion mass M_1 for various ion fluences at 30 keV ion energy. The solid line represents a $M_1^{-1/2}$ dependence [6.23]

fringes corresponding to the graphite layers have completely disappeared and the texture has become like that of amorphous carbon. These changes in the graphite layer are also reflected in the optical diffractograms shown in the insets of Fig. 6.7. As the fluence increases, each sharp spot in the optical diffractograms of the pristine fiber develops into a collection of more diffuse spots indicating that implantation to high fluences causes a large decrease in parallelism of the layers with respect to the fiber axis. Thus by increasing the fluence, not only are the crystallite sizes reduced but also the parallel arrangement of the crystallites with respect to the fiber axis is also reduced.

An independent measurement of the implantation-induced reduction in L_a which is consistent with the TEM results for ion implanted graphite was obtained by *Elman* et al. from I_D/I_G the ratio of the intensity of the disorder-induced Raman line at 1360 cm^{-1} to that of the Raman-allowed line at 1582 cm^{-1} (Sect. 5.1 and Fig. 6.2) [6.4]. Figure 6.8 shows the dependence of the in-plane (L_a) crystallite size on implant ion mass for several implantation fluences at 30 keV ion energy [6.23], based on the measurements of (002) lattice images. L_a and L_c were also determined from the (002) dark field images by measuring the lengths of the bright 'spots' along the directions parallel and perpendicular to the fiber axis, respectively. The crystallite sizes obtained from the (002) dark field and lattice images were found to give a similar dependence of the crystallite size upon fluence and ion mass. As indicated in Fig. 6.8, the dependence of the crystallite size on ion mass M_1, follows approximately a $M_1^{-1/2}$ dependence for a fluence of 1×10^{15} ions/cm^2, whereas a weaker dependence on the ion mass was found at lower fluences.

The c-axis interplanar spacings were obtained by taking optical diffractograms from the negatives of the lattice image micrographs and the results are shown as insets to Fig. 6.7a-d. Analysis of the optical diffractograms indicates an increase in the c-axis interplanar spacing after implantation. Interplanar distances up to 3.53 Å have also been obtained from measurements of dimensional changes on pyrolytic graphite irradiated with neutrons ($\phi \sim 2 \times 10^{21} n/\mathrm{cm}^2$) at high temperatures ($> 450°$C) [6.24]. The increase in interlayer spacing is larger for heavy ions than for lighter ones, which suggests that in part the ions may come to rest in interstitial positions between the graphite layers. Interplanar spacings as high as 3.44 Å have been calculated from X-ray diffraction patterns for carbon samples graphitized at temperatures ($\sim$ 1600°C) where C atoms are known to reside interstitially [6.24]. Furthermore, vacancy clusters have been observed by *Kelly* et al. in natural graphite flakes irradiated with neutrons ($\phi \geq 10^{16}/\mathrm{cm}^2$) and C ions ($\phi \geq 10^{16}/\mathrm{cm}^2$) at high temperatures [6.24]. The detailed nature of the defect sites has not yet been fully elucidated.

Further insight into what is happening in graphite implanted at the higher fluence levels, where for example the RBS, STM and Raman spectra all indicate complete amorphization, is obtained by other characterization techniques which are more sensitive in the high fluence range. For example, scanning electron microscopy (SEM) studies (Fig. 6.9) show that the onset fluence where the Raman spectra rapidly broaden corresponds to the appearance of surface damage on the SEM electron micrographs, first as visible ridges, which at higher ion doses (Fig. 6.9) split open, exposing fractured graphite layers on the surface [6.2]. The surface ridges and the surface fracture may be attributed to the relief of implantation-induced local stress and strain. Furthermore, at these high dose levels, blisters containing a high concentration of Ar gas are formed, and these blisters are then ruptured to produce surface fracture.

A more detailed study of this aspect of surface damage due to ion implantation has been carried out by *Yugo* et al. [6.25]. By using 150 keV Ar^+ ions at a beam current density of 3 $\mu\mathrm{A/cm}^2$ for room temperature implantation and covering a dosage range of $10^{13} - 10^{17}$ ions/cm^2, *Yugo* et al. showed that surface tears appear at a dpa value near unity (Fig. 6.9). By considering the expression for the critical fracture stress

$$\sigma_c = \left(\frac{4Y\gamma}{\pi l_c}\right)^{1/2} \tag{6.1}$$

to form a crack of length l_c (which was measured to be in the 10–100 μm range [6.25]), *Yugo* et al. estimated the effective modulus Y in (6.1), using an effective surface energy γ in the $10^3 - 10^5$ erg/cm^2 range [6.25]. From their analysis, *Yugo* et al. found that Y was in the $10^6 - 10^8$ dyne/cm^2 range, characteristic of a modulus value for grain boundaries given by *Soule* and *Nezbeda* [6.26], but not for crystalline graphite for which Y is in the $10^9 - 10^{10}$ dyne/cm^2 range [6.24]. On this basis, it was concluded in [6.25] that the cracks are initiated at a grain boundary and that the cracks propagate along grain boundaries. As they propagate, the cracks join other cracks to form polygons when the crack length l_c is comparable to the grain boundary dimensions [6.25]. Blister formation was also observed [6.25] at high fluences ($\sim 10^{17}/\mathrm{cm}^2$) for rare gas ion implants, and the blistering was attributed to the formation of tiny gas bubbles, which coalesce

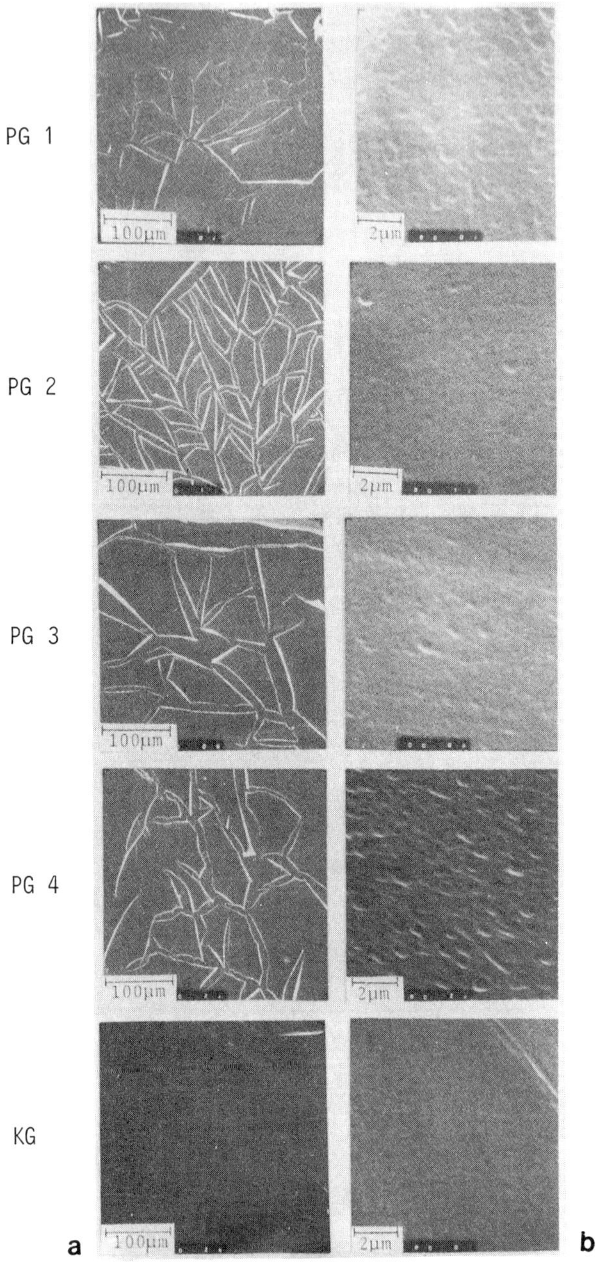

Fig. 6.9. Scanning electron micrographs by *Yugo* et al. of various graphite samples implanted with 150 keV Ar^+ ions: The left column, (**a**) shows cracks produced at a dose of 3×10^{15} ions/cm^2 and the right column, (**b**) shows blisters formed at 3×10^{17} ions/cm^2. Note the differences in magnification [6.25]. [PG: pyrolytic graphite, KG: kish graphite (Sect. 2.1.2)]

to form larger bubbles. Some of these larger bubbles propagate to the surface along non-basal screw dislocations within single grains [6.27]. To support their interpretations, *Yugo* et al. [6.25] carried out identical SEM measurements on various pyrolytic and kish graphite samples (see Sect. 2.1.2 for an explanation of pyrolytic and kish graphites) and their results show greatly reduced crack and blister formation in kish graphite, which has much larger grains and a much lower non-basal dislocation density ($10^5/\mathrm{cm}^2$ as compared with $10^6/\mathrm{cm}^2$ for HOPG).

Complementary information is obtained by Mössbauer spectroscopy and by scanning transmission electron microscopy (STEM). For ion doses high enough to lead to RBS and Raman spectra similar to those for an amorphized near-surface region, measurements by *Schroyen* et al. of the angular dependence of the recoilless fraction using the Mössbauer technique show that an anisotropic microstructure remains even after extensive lattice damage had been introduced by 85 keV ^{57}Co or ^{133}Xe ions implanted to a dose of 10^{14} ions/cm^2 [6.28]. Extensive studies of the anisotropic nature of the ion-beam induced amorphized microstructure were performed by *Söder* et al. by probing the anisotropy of the self-diffusion in ^{13}C implanted HOPG [6.29]. The findings of these studies are consistent with STEM studies on an HOPG sample implanted with 100 keV ^{75}As ions to a dose of 10^{15} ions/cm^2. The STEM studies show, in addition to very heavily damaged regions, some undamaged regions (as large as $\sim$100Å) exhibiting selected area diffraction patterns characteristic of microcrystalline graphite [6.30]. The presence of these microcrystalline regions are characteristic of disordered carbon produced by ion implantation, and they can serve as seeds for crystalline regrowth under appropriate annealing conditions.

In this section the present knowledge of the lattice damage introduced by ion implantation into graphite has been reviewed. Since most of these ion implantation studies to date have addressed the lattice damage issue, much information has been accumulated, covering a wide range of experimental techniques, including RBS, Raman and Mössbauer spectroscopy, TEM, SEM, STEM, STM, SEE, X-ray diffraction, fracture studies and others. All of these studies yield complementary information and show increased lattice damage with increased ion dose and ion mass, up to levels of implantation where amorphization sets in (corresponding to about 1–2 dpa). Further ion implantation has a reduced effect on modifying the properties of the graphite host material.

The reason why so much of the literature has focused on the lattice damage issue is that disordered graphites are prevalent in most commercially useful graphite-related materials. However these commercial disordered graphites tend to be less homogeneous in texture and less controllable in processing than ion-implanted graphite. For this reason, ion-implanted graphite provides a model system for the production of disorder in graphite, as is also emphasized in the introduction to this section. The following section reviews the current state of knowledge on the graphitization of ion-implanted graphite, another topic that dominates the literature on ion-implanted graphite because of the scientific and commercial importance of the graphitization process.

6.2 Regrowth of Ion-Implanted Graphite

As discussed in detail in Sect. 6.1, ion implantation provides a very convenient method for introducing disorder into the near-surface region of graphite in a controlled and reproducible manner, thereby allowing for the systematic study of recrystallization phenomena and of the graphitization process in this anisotropic material. Regrowth and graphitization studies following ion implantation of graphite are summarized in this section.

Once again the complementary information provided by Raman spectroscopy, RBS, and TEM has been exploited to obtain a comprehensive characterization of the recrystallization process both in the basal planes and along the c-axis direction of ion-implanted graphite. These complementary studies further provide a detailed understanding of the evolution of the microstructure and regrowth kinetics in both directions of this highly anisotropic material. The recrystallization studies of graphite following ion irradiation are closely related to studies of graphitization by *Oberlin* and coworkers by heat treatment of various precursor carbons [6.31].

Carbons are classified into graphitizable and ungraphitizable precursors. The graphitizable carbons generally start with either large or small ordered planar basic structural units (a few layers thick) which are aligned to some extent. In the graphitization process, these planes grow laterally, and the mosaic spread of their c-axis orientations decreases until they obtain the AB registry of three-dimensional graphite. The ungraphitizable carbons generally have precursor morphologies which could best be described as tangled ribbons of graphite-like material. Upon heat treatment, these more-or-less planar ribbons grow but, because of constraints imposed by the original tangle, they often cannot fully form three-dimensional graphite. The conventional wisdom about the characteristic temperatures associated with the graphitization process is as follows. The carbon atoms in the disordered (amorphous) region migrate to the edges of the crystallites on the ribbons and form trigonal bonds with the planar edge atoms; this process starts at about 1500°C. The energy required for this process (~0.4 eV) is much less than that associated with the σ-bond so that the rate limiting step for the graphitization process may possibly be associated with carbon atom migration. In the temperature regime between 1500° and 2300°C the separation between the layer planes in graphitizable carbons is observed to decrease from ~3.45Å down to about 3.35Å. The initial decrease in interplanar separation to 3.42 Å is associated with plane flattening of the ribbons, which must occur when the crystalline platelets increase in area, but the graphene planes still are turbostratically decoupled from adjacent graphene layers. Finally as the temperature reaches ~2500°C, where the crystallite sizes as inferred by X-ray and Raman measurements are greater than 300Å, the three-dimensional AB stacking order sets in and larger three-dimensional crystals are formed. The three-dimensional registration provides the closest packing of the layer planes with the minimal c-axis atomic separation of 3.35Å. This is reached for heat treatment temperatures $T \geq 3000°C$ [6.3].

A variety of mechanisms appear to be operative in the restoration of crystalline order upon the annealing of ion-implanted graphite, making it difficult to provide a unified description of these processes. Two main annealing stages have been identified in ion-damaged graphite. The first stage involves in-plane or 2D ordering and takes place for heat treatment temperatures in the range $300°C \le T_a \le 2300°C$, yielding a turbostratic material in which the graphene layers lack 3D interplanar registration. This is followed by a second graphitization stage above ~2300°C where, in addition, three-dimensional ordering takes place. However, other regrowth modes have also been identified like random crystallization, heterogeneous crystallization at residual crystallites and epitaxial regrowth. The occurrence of one or another of these processes depends strongly on the initial amount of damage in the lattice. Therefore the initial amount of disorder in the graphite structure provides a convenient framework for reviewing the recrystallization studies in ion-implanted graphite.

A detailed study of the regrowth of ion-implanted graphite was carried out with carbon ion implants [6.3, 32] to eliminate complications in the regrowth arising from the introduction of impurities. For comparison, similar measurements were also made with impurity implants such as ^{31}P and ^{75}As. The most comprehensive information about the regrowth process has been provided by RBS/channeling measurements. RBS spectra by *Venkatesan* et al. illustrating this regrowth process are shown in Fig. 6.10 for ^{12}C ions implanted into HOPG at 230 keV to a fluence of 10^{15} ions/cm^2. This fluence is sufficiently low so that some ion channeling can still take place in the near surface region. Figure 6.10 shows RBS spectra taken with 2.0 MeV ^{4}He ions for isochronal anneals (t_a = 20 min) for a variety of annealing temperatures T_a. At low T_a (< 1000°C), regrowth occurs epitaxially as indicated by the advance of the order-disorder interface towards the surface and by heterogeneous annealing of the region close to the surface, where small crystallites are available as needed for crystal regrowth. For annealing temperatures of 1000°C and above, annealing also occurs in the regions experiencing the maximum lattice damage upon implantation (Fig. 6.1). With increasing T_a, this lattice damage is further healed and at an annealing temperature of ~ 2300°C, the regrowth shown by the RBS channeling data of Fig. 6.10 is essentially complete, in agreement with Raman studies on ion-implanted HOPG [6.32]. Interestingly, the temperature of 2300°C corresponds to the completion of 2D ordering of the graphene planes and the onset of 3D ordering in the graphitization process, as studied in soft carbons by other means, such as high resolution transmission electron microscopy and reviewed by *Oberlin* [6.31].

The annealing of ion-irradiated graphite has also been studied by *Kazumata* et al. using the ESR technique, showing that annealing of graphite implanted with Ar^+ (1.1×10^{14}/cm^2 at 450 keV), N^+(1.1×10^{15}/cm^2 at 1.1 MeV), and He^+(1.3×10^{15} and 1.3×10^{16}/cm^2 at 400 keV) at a temperatures T_a below 1000 K produced no changes in the magnitude of $g_{\|}$ nor in $g_{\perp}$ [6.21]. However, the intensity of the ESR line decreased gradually with increasing T_a. The temperature dependence of the ESR signal could be explained in terms of the Curie

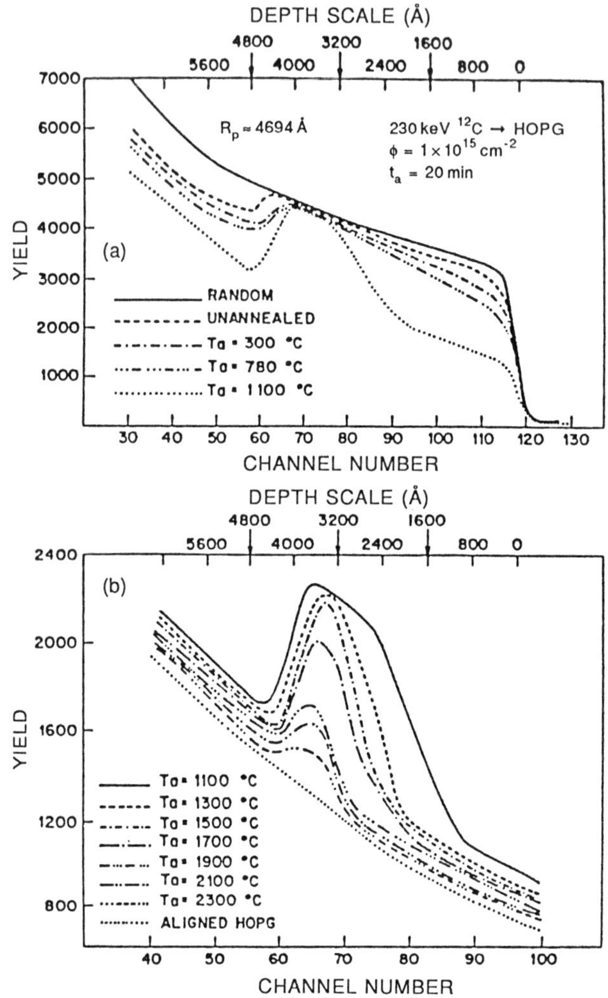

Fig. 6.10. Regrowth of the partially disordered region of HOPG implanted with 230-keV ^{12}C ions at a fluence of $\phi = 1 \times 10^{15}\ cm^{-2}$ anneal for 20 min at various temperatures in the range (**a**) $300 < T_a < 1100°C$ and (**b**) $1100 < T_a < 2300°C$ as observed by *Venkatesan* et al. using the RBS channeling technique [6.3]

Law, indicating that the main mechanism for the ESR absorption was due to localized spins [6.21]. Thus, the results indicated a decrease in the concentration of unpaired spins, with few unpaired spins remaining for $T_a > 1000$ K. These results for ion-implanted graphite are in contrast to those for neutron-irradiated graphite, where the lattice damage is predominantly associated with point defects that can be annealed out by heating to 1300 K [6.33]. The difference in the damage mechanism in these two cases may well have to do with the fact

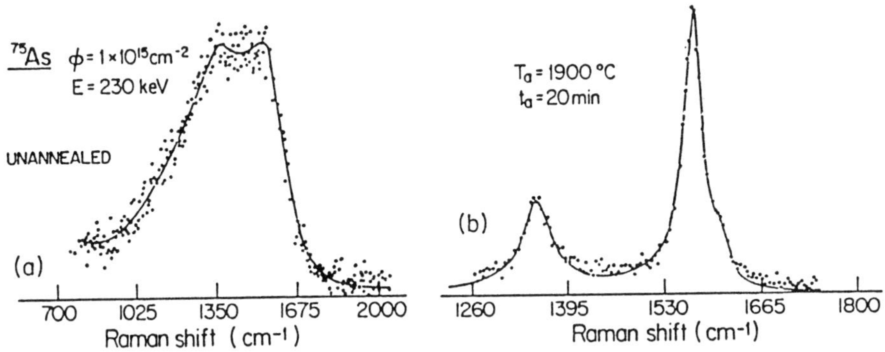

Fig. 6.11. Raman spectrum of HOPG ion-implanted with ^{75}As at an energy of 230 keV and a fluence of 1×10^{15}cm^{-2} (**a**) before and (**b**) after annealing at 1900°C for 20 min [6.1]

that heavy ion implantation leads to a dense damage cascade which neutron irradiation does not.

These observations suggest that the type of damage which is annealed out at $T_a \leq 2300$°C is mostly due to interstitials and is likely mediated by their diffusion. The preceding assumptions are supported by the results of the analysis of RBS studies in terms of Arrhenius plots of the movement of the order-disorder interface in the epitaxial regrowth of ion-implanted graphite. Arrhenius plots of the number of displaced carbon atoms (area under the disorder peak in the RBS channeling spectra of Fig. 6.1) were made upon annealing the implanted samples over a wide range of temperatures, for both carbon implants and impurity (e.g., ^{75}As) implants. The results for the T_a dependence of the density of displaced carbon atoms yield a very low activation energy of 0.15 eV which is identified as the energy required by an interstitial to overcome its potential barrier, as it moves from a site that blocks the (001) channel to a lattice site [6.3]. The energy required for vacancy motion is very much greater, ~ 2.0 eV for the activation energy for vacancy motion in the basal plane, and ~ 5.5 eV for vacancy motion along the c-axis [6.34].

Furthermore, studies by *Venkatesan* et al. of the RBS-channeling spectra for different annealing times have provided information about the kinetics of the regrowth process [6.3]. Initially, the regrowth process has been shown to follow a $t_a^{1/2}$ time dependence, consistent with a diffusion-limited process for the migration of the interstitial atoms to lattice sites [6.3]. Figure 6.11 shows the Raman spectra of HOPG after implantation with ^{75}As ions at an energy of 230 keV and a fluence of $1{\times}10^{15}$/cm^2 (Fig. 6.11a) and after subsequent annealing at 1900°C for 20 minutes [6.1]. The Raman spectrum of the as-implanted sample exhibits a very broad band consistent with the presence of a heavily disordered structure in the near-surface region. Upon annealing, three lines are resolved. The strongest line, peaked at about 1580 cm^{-1}, corresponds to the Raman-allowed zone-center high-frequency mode of HOPG. The other two are the disorder-induced lines appearing because of the high density of phonons

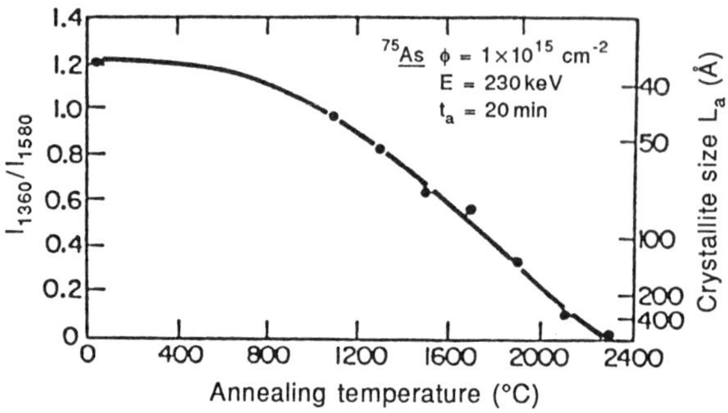

Fig. 6.12. Ratio of Raman scattering intensities of the disorder-induced D line at $\sim$ 1360 cm^{-1} to the Raman-allowed G line at $\sim$ 1580 cm^{-1} vs annealing temperature T_a for samples of HOPG implanted with ^{75}As ions under the same conditions as in Fig. 6.11, and annealed for 20 min (isochronal annealing) at various annealing temperatures [6.1]

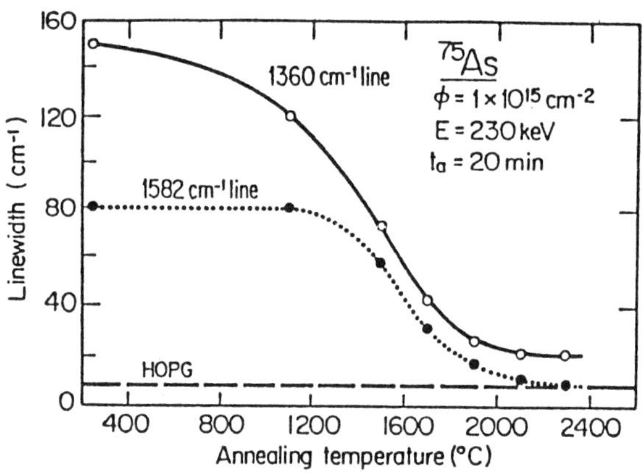

Fig. 6.13. Plot of the line width (FWHM) of the disorder-induced D line at $\sim$ 1360 cm^{-1} and of the Raman-allowed G line at 1582 cm^{-1} vs annealing temperature T_a, corresponding to the same spectra as in Fig. 6.12. The width of the Raman line for unimplanted HOPG is indicated by a heavy dashed line labeled HOPG [6.1]

near $\sim$1360 cm^{-1} and 1620 cm^{-1} [6.1]. As described previously in Sect. 5.1, the ratio of the integrated intensities of the disorder-induced $\sim$1360 cm^{-1} line to the Raman-allowed $\sim$1580 cm^{-1} G line ($R = I_D/I_G$) provides a very convenient estimate of the in-plane crystalline size. The disorder-induced line at $\sim$1620 cm^{-1} appears as a shoulder on the G line in Fig. 6.11b. Analogous information can be obtained from measurement of the line widths of the corresponding Raman peaks which are inversely proportional to the crystallite size in the disordered material.

As shown in Figs. 6.12 and 6.13, both the Raman intensity ratio R and the Raman line widths decrease with increasing annealing temperature. After

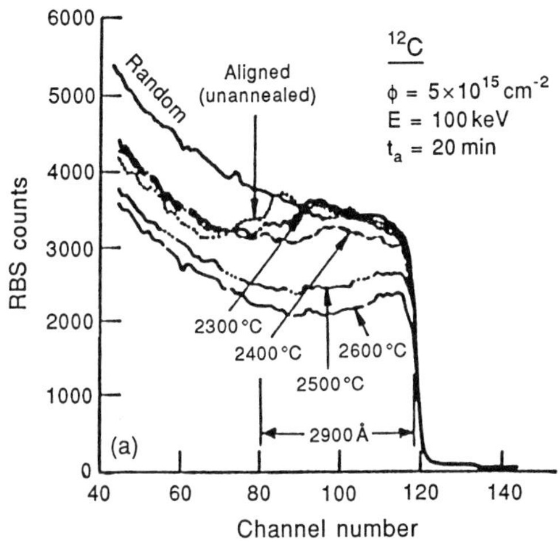

Fig. 6.14. Random and aligned RBS spectra reported by *Elman* et al. for several annealing temperatures for HOPG ion-implanted with ^{12}C ions at an energy of 100 keV and a fluence of 5×10^{15}cm^{-2}. The annealing time at the indicated temperatures was 20 min [6.1]

annealing at 2300 °C the disorder-induced lines totally disappear and the width of the Raman-allowed line returns to its value in HOPG prior to implantation. Surprisingly, the aligned RBS channeling spectra of the samples annealed at 2300 °C, still coincide with the random spectrum in the implanted region.

This apparent paradox was solved with the help of TEM measurements which confirmed that the near-surface region probed by all three techniques (RBS, Raman and TEM), consisted of 2D planes of graphite after annealing at 2300 °C. The TEM studies however showed that these turbostratic graphene layers had no correlation in the third (c-axis) direction, to which RBS-channeling measurements are sensitive. These results are in good agreement with the temperature for establishment of 2D ordering in the graphitization process of a variety of disordered carbons [6.31]. Ordering of implanted graphite along the c-axis occurs at annealing temperatures higher than 2300 °C and proceeds through a different mechanism. In this connection, Fig. 6.14 shows channeling spectra for isochronal annealing (t_a=20 minutes) of HOPG implanted with ^{12}C at $\phi = 5\times10^{15}$cm^{-2} and E = 100keV over a wide range of annealing temperatures ($T_a < 3000$°C). The "random" and aligned channeling spectra for HOPG and the spectrum for the as-implanted (unannealed) sample are given for comparison. For annealing temperatures in the range $T_a < 2300$°C the regrowth is mainly epitaxial, moving the order-disorder interface closer to the surface.

In contrast, at higher annealing temperatures ($T_a > 2300$°C), the backscattering yield decreases concomitant with the advancement of the interface. This implies one of two possibilities: either the formation of well-ordered two-dimensional regions with partial three-dimensional registry, or the formation of

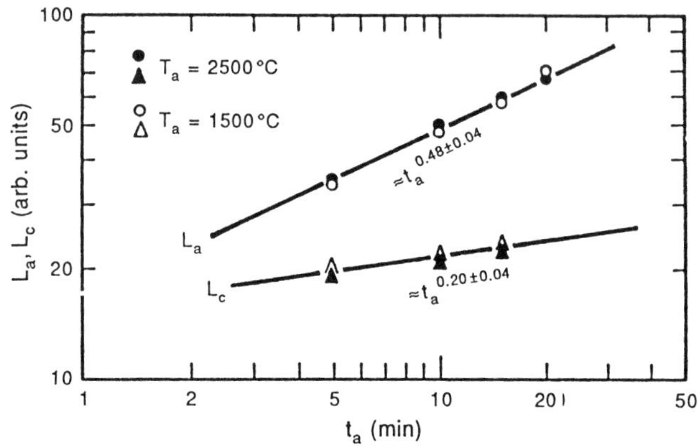

Fig. 6.15. In-plane (L_a) and c-axis (L_c) crystallite sizes versus annealing time t_a for HOPG samples which had previously been implanted at room temperature with 30 keV ^{209}Bi to a fluence of $1\times10^{15}/\mathrm{cm}^2$. After the implantation at room temperature the samples were annealed at 1500°C and 2500°C for various annealing times [6.22]

three-dimensionally-ordered islands in a disordered host. In both cases, however, there is a decreasing registry of the layers when moving from the undamaged bulk region towards the surface. This implies that independent of which of the above scenarios is correct, the three-dimensional registry occurs in an epitaxial manner.

The 3D ordering of the layer planes takes place, with a relatively high activation energy of ~ 1.2 eV [6.3]. In the regime of high initial ion fluence (5×10^{15} ions/cm^2 for 100 keV ^{12}C ions and $T_a < 2300$°C), electron diffraction patterns show (001) diffraction rings from a nominal c-face surface, clearly indicating the presence of randomly-oriented crystallites. This implies that heterogeneous nucleation and regrowth take place in addition to the epitaxial regrowth at the order-disorder interface [6.22] for graphite exposed to ion implantation at lower fluences.

More detailed information on the regrowth kinetics has been provided by high resolution transmission electron microscopy (HRTEM) work of *Salamanca-Riba* et al. (Sect. 5.3) on both ion-implanted HOPG and vapor-grown carbon fibers heat treated to temperatures between 2900°C and 3500°C; these allowed independent examinations of the kinetics of the in-plane and c-axis regrowth. To maximize the sensitivity of the TEM technique, shallow implantations with 30 keV ^{209}Bi ions were done [6.22]. Measurements of both the in-plane and c-axis correlation lengths L_a and L_c were carried out as a function of both annealing time and annealing temperature and it was shown that at both low temperature (1500°C) and high temperature (2500°C), the in-plane regrowth followed a $t^{1/2}$ dependence, consistent with a diffusion mechanism (Fig. 6.15).

Kinetic studies using the RBS technique showed that the regrowth of the more heavily damaged regions follows a slower and more complicated time dependence than $t^{1/2}$ [6.3]. As shown in the Fig. 6.15, the c-axis regrowth follows

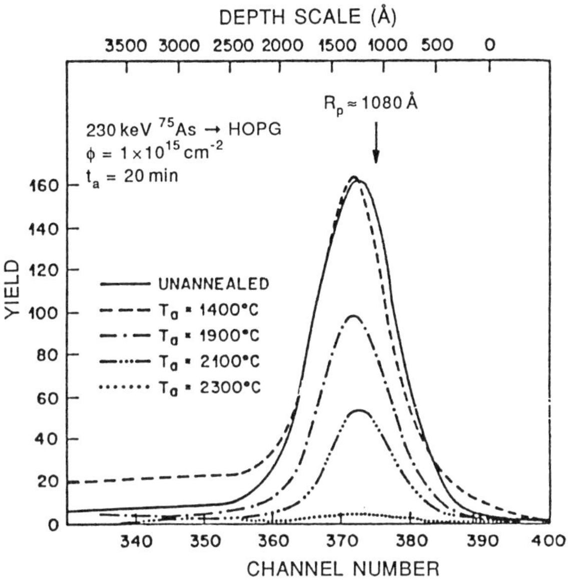

Fig. 6.16. RBS depth distribution profiles of arsenic ions implanted in HOPG at $E =$ 230 keV and $\phi = 1\times10^{15}$ cm^{-2} before and after isochronal annealing ($t_a = 20$ min) at various temperatures in the range $1400° < T_a < 2300°$C. The curve corresponding to the unannealed sample is essentially indistinguishable from the curve for the sample annealed at 1100°C [6.3]

(within the experimental uncertainty) a $t^{1/4}$ time dependence, consistent with the climb of dislocations in the c-axis regrowth process [6.22].

As the regrowth occurs, impurities are expelled from the graphite structure. Since the RBS spectra allow independent study of the impurity profile (Sect. 4.1), the expulsion of the impurities during the regrowth process can be studied in detail. Figure 6.16 shows the reduction in the impurity concentration as T_a is increased; in this figure specific data by *Elman* et al. are presented for the case of ^{75}As implanted into HOPG at 230 keV to a dose of 10^{15} ions/cm^2 [6.3]. The absence of any channeling effect for the arsenic RBS peak indicates that the impurities in graphite do not occupy lattice sites, but rather are located interstitially, consistent with studies using other characterization techniques, such as X-ray diffraction which show a significant increase in the interlayer separation as a result of ion implantation [6.35]. The absence of any change in lineshape in the profiles of Fig. 6.16 as T_a is increased (all the way up to 2300°C) implies the absence of diffusion along the c-axis. However, the As ions in Fig. 6.16 show a decrease in FWHM linewidth with increasing T_a, indicating that at higher temperatures the impurities are relatively better retained in the most heavily damaged regions of the sample. By plotting the amount of As that is released from the sample at a given annealing temperature in an Arrhenius plot, the activation energy for the As atoms leaving the sample can be obtained, yielding

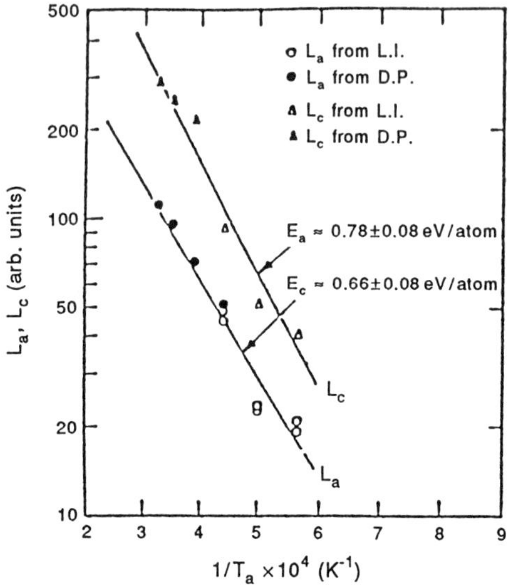

Fig. 6.17. Arrhenius plot of in-plane (L_a) and c-axis (L_c) crystallite sizes versus reciprocal annealing temperature $1/T_a$ measured by *Salamanca-Riba* et al. from HRTEM lattice images (L.I.) and diffraction patterns (D.P.) for carbon fibers previously implanted with ^{75}As at 230 keV to a dose of 1×10^{15}cm^{-2} and subsequently annealed for 1 h [6.22]

~ 0.75 eV, in good agreement with that found for in-plane crystallite growth (0.78 eV) and for c-axis crystallite growth (0.66 eV) as determined by TEM measurements (Fig. 6.17) of L_a and L_c from the lattice fringe images (L.I.) and the electron diffraction patterns (D.P.) [6.22].

The expulsion of impurities facilitates 3D crystal regrowth, which also results in a reduction in the c-axis lattice constant to approximately the single crystal value [6.22, 35]. RBS studies of the impurity profile by *Elman* et al. show that the impurities with large ionic radii are expelled at relatively lower annealing temperatures T_a than those with small ionic radii (Fig. 6.18) [6.36], consistent with the squeezing out of impurity ions through crystallite edges during regrowth. For room temperature implantation the expulsion of impurity ions is essentially complete for $T_a \sim 2300$°C. However, implantation of charged impurities at elevated temperatures (≥ 600°C) has been shown to enhance the retention of the implant, even after subsequent annealing to temperatures as high as 2300°C [6.36].

It can thus be concluded that the regrowth of the graphite, which was previously disordered through ion implantation, is similar to the graphitization process itself, showing in-plane crystallite growth to ~ 100Å followed by the introduction of three-dimensional stacking order [6.31]. However, the regrowth process for the implanted samples differs through the presence of a well-defined order-disorder interface across which there is a discontinuity in the values of L_a, L_c and of the interplanar separation. This interface also plays an important

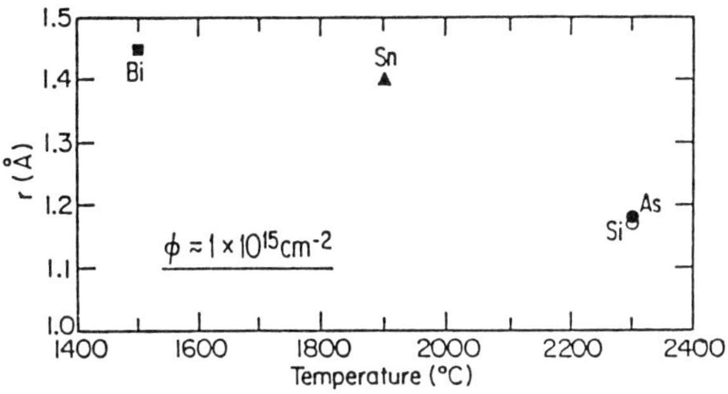

Fig. 6.18. The relationship between the annealing temperature at which practically no implanted species is left in HOPG samples and the covalent radii of the implanted ions for various implants into graphite at a fluence of 1×10^{15} cm^{-2} [6.36]

role by promoting epitaxial regrowth from the interface and eventually growing all the way to the surface. Apart from this, epitaxial regrowth on small crystallites, heterogeneous nucleation and regrowth, and the expulsion of impurities are all features common to both the regrowth of ion-implanted graphite and graphitization in general.

6.3 Structural Modification

Normally, ion implantation causes a lattice expansion of the host material as interstitials are formed, thereby introducing high local stresses that are relieved by lattice expansion. Because of its high degree of anisotropy, graphite is a unique material with regard to implantation-induced modification of its structure. For example, neutron-induced lattice damage causes a lattice expansion along the *c*-axis [6.37], to accommodate the stresses introduced by the interstitials (as also occur in other materials), but a lattice contraction in the basal plane to accommodate the stresses introduced by irradiation-induced vacancy formation [6.24, 38]. The reason for this exceptional behavior is related to the large anisotropy and very small intraplanar C–C distance (Sect. 2.1.1), which inhibit interstitial formation within the basal plane. On the other hand, the weak interplanar coupling allows for an easy accommodation of interstitials between the graphene layers, as suggested by *Kelly* [6.24]. Figure 6.19 shows a plot of the in-plane and *c*-axis lattice constants as a function of neutron fluence [6.24]. No detailed lattice constant measurements have yet been reported for ion-implanted graphite.

However, some measurements pertinent to ion-induced structural changes have been carried out in ion-implanted graphite. By taking the Fourier transform of the TEM lattice fringes in ion-implanted graphite [6.35], two dominant effects were observed: firstly, a broadening of all diffraction spots (strongly correlated with implantation-induced lattice damage), as well as an implantation-

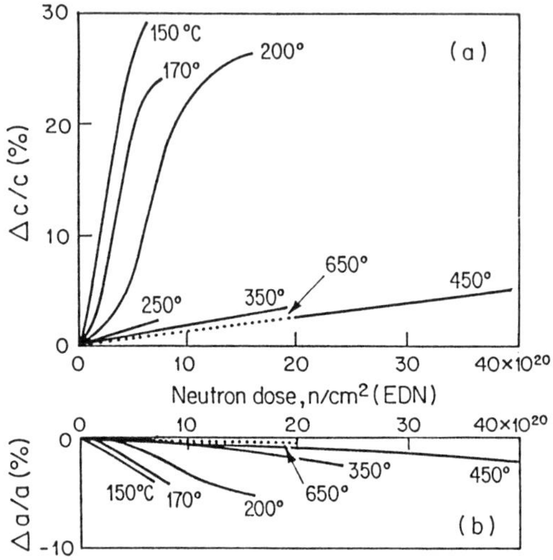

Fig. 6.19. Plot of increase in (**a**) c-axis and (**b**) in-plane lattice constant of neutron-irradiated HOPG as a function of neutron fluence for various irradiation temperatures [6.24, 37]

induced expansion of the c-axis interplanar separation. It was further found [6.35] that this c-axis lattice expansion was not readily restored to its pre-implantation value, even after annealing to 1000°C, presumably because of the presence of interstitials. More detailed studies showed this retention of implantation-induced impurities at 1000°C to be related to the graphitization process itself, with eventual expulsion of the interstitials by 2300°C [6.36] (Sect. 6.2).

Qualitative information on the effect of ion-implantation on the lattice structure at very low ion doses comes also from studies by *Elman* and coworkers of the implantation-induced changes in the electronic structure [6.2, 39], showing a small decrease in the intraplanar nearest-neighbor overlap integral γ_0, and a larger decrease in the interplanar nearest-neighbor overlap integral γ_1 (Sect. 6.4). The small decrease in γ_0 implies a small increase in the in plane lattice constant a, inconsistent with the *Kelly* model for neutron-irradiated graphite with regard to the sign of the change in a. However, the larger decrease in γ_1, relative to γ_0 implies a larger increase in c_0 relative to the increase in a, consistent with the anisotropy of the bonding forces in graphite and also consistent with the *Kelly* model for neutron-irradiated graphite [6.24].

One possible explanation for the difference between neutron irradiation and ion implantation is that no new ion species are introduced by neutron irradiation (the neutrons are not massive and they do not come to rest inside the target), so that the number of vacancies is equal to the number of interstitials. In contrast, ion implantation locally increases the atomic density by the introduction of new atomic species into the lattice, so that the net stress-induced expansion is expected to be greater. Thus, in comparing neutron irradiation with ion

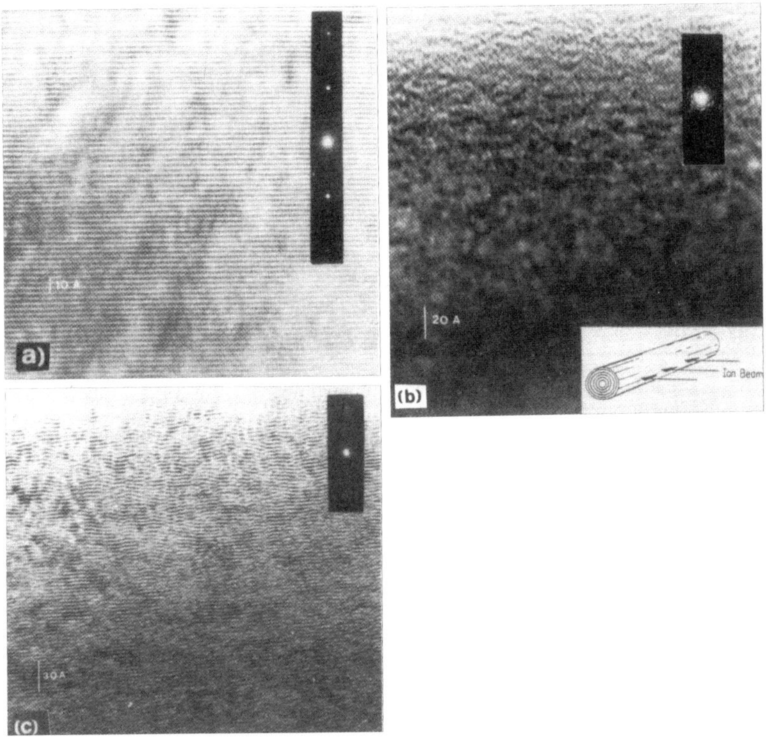

Fig. 6.20. HRTEM (002) lattice images of (**a**) a pristine vapor-grown fiber heat treated to 3000°C, (**b**) a similar fiber implanted with ^{209}Bi ions to a dose of 1×10^{15} ions/cm^2 (together with a schematic representation of the ion implantation configuration), and (**c**) an implanted fiber [as in (**b**)] after annealing at 1500°C for 1 h. The insets to (**a**), (**b**) and (**c**) are optical diffractograms taken from the negatives of the lattice images. From the lattice fringe images, L_a and L_c are determined as a function of annealing time t_a and annealing temperature T_a [6.22]

implantation, a much smaller volume expansion would be expected in the case of neutron irradiation. A Poisson ratio argument could then perhaps explain why an in-plane contraction might accompany a c-axis expansion for neutron irradiation, whereas the larger lattice expansion for ion implantation would give rise to a lattice expansion in all directions. The fact, that ion implantation creates dense damage cascades which may leave parts of the material vacancy-rich, and other parts interstitial-rich, is another possible reason for the excessive swelling observed in graphite [and also in diamond (Sect. 8.2)] as a result of heavy ion implantation.

In correlating the increased wear resistance associated with ion implantation to the modification of other properties, *Kenny* et al. [6.40] reported a shrinkage of the implanted region at high ion dose levels, observed through optical and SEM experiments [6.40], though no detailed measurements were presented on the effect of the implantation on the lattice constants.

The most dramatic implantation-induced change in the graphite structure relates to damage-induced changes in the microstructure. Low doses of low mass ions give rise to point defect formation and a radical reduction of the in-plane and c-axis crystallite sizes, L_a and L_c, respectively. Implants with higher masses and higher doses cause more extended defects, eventually leading to amorphization. The most extensive study of the implantation-induced changes in the microstructure has been carried out by *Salamanca-Riba* et al. using high resolution transmission electron microscopy [6.22]. In this work, heavy ions (^{209}Bi) at low energies (30 keV) were used to confine the implanted species and the associated lattice damage to the near-surface region, to better match the ion penetration depth to the length scale over which the TEM measurements are sensitive. Figure 6.20 shows an example of the great reduction in the crystallite size that is induced by a dose of $10^{15}/\text{cm}^2$ ^{209}Bi ions at 30 keV, and the subsequent regrowth (Sect. 6.2) after partial annealing of the lattice damage.

6.4 Modification of the Electronic Structure and Transport Properties

Studies of the effect of ion implantation on both the electronic structure and the transport properties of graphite have been carried out, using many of the same techniques that have been found to be especially sensitive to these properties in the reference crystalline state. The general trends observed in the modification of these properties are briefly described in this section.

The electronic structure for single-crystal graphite is well established [6.41]. The most sensitive technique that has been used for determination of the electronic energy dispersion relations for graphite is the magnetoreflection technique. Because of its high sensitivity, this technique was applied by *Elman* et al. to the study of ion implantation-induced changes in the magnetoreflection spectrum of graphite [6.39]. Since the appropriate frequency range for magnetoreflection studies is in the infrared, the penetration depth for light is approximately an order of magnitude greater than for the ion beam. Nevertheless, changes in the magnetoreflection spectra could be observed. The measurements were carried out on highly oriented pyrolytic graphite (HOPG) implanted with ^{31}P ions at 600°C in the energy range $70 < E < 200$ keV, and for fluences in the range $8.5 \times 10^{13} < \phi < 1.0 \times 10^{15}$ cm^{-2}. The implantation was carried out at elevated temperatures (600°C) to reduce the lattice damage in the region traversed by the implanted ions, and to maintain a smooth optical surface. Subsequent to implantation, the samples were characterized by Raman spectroscopy to measure the disorder in the optical skin depth, by SIMS (secondary ion mass spectrometry) to determine the implantation profile in the near surface region, and zero field infrared reflectivity was used to provide a convenient reference for the magneto-optic experiments [6.39].

The magnetoreflection experiments showed that the spectra could be fit by the same Slonczewski-Weiss-McClure (SWMcC) energy band structure as is used to describe the energy dispersion relation of crystalline graphite within an electron volt of the Fermi level, only with slightly modified values for the band parameters of the SWMcC model. By detailed study of the magnetoreflection spectra for interband transitions at the H-point and at the K-point along the edge of the hexagonal Brillouin zone (Fig. 2.11), it was possible to determine the implantation-induced modification to both γ_0 and γ_1, the in-plane and c-axis nearest-neighbor overlap integrals. The magnetoreflection results show that both γ_0 and γ_1 decrease with increasing ion fluence (i.e., increasing lattice damage), consistent with a lattice expansion in all directions. However, the decrease in γ_1 (and consequently in c-axis lattice expansion) was found to be much larger than that in the basal plane, consistent with the anisotropic binding in graphite [6.39]. By making measurements as a function of ion energy, the ion penetration depth was varied from 670 Å to 1980 Å, and a sharp decrease in magnetoreflection signal was observed with increasing ion energy. This observation is consistent with the increased fraction of the magneto-optically probed skin depth that had experienced significant lattice damage. Increasing the photon energy $\hbar\omega$ results in a decrease in the skin depth for the infrared radiation, so that the implantation-damaged region in this case also corresponds to a larger fraction of the skin depth for the IR radiation.

The band overlap in graphite is almost completely determined by the band parameter γ_2 of the graphite electronic structure (Sect. 2.2.2) which depends on weak interactions between atoms two layers apart. Since γ_1 depends on interactions between adjacent layers, it follows that $(\Delta\gamma_2/\gamma_2) \approx 2(\Delta\gamma_1/\gamma_1)$ due to changes in band structure that are induced by ion implantation. These arguments imply that ion implantation should decrease the band overlap, and decrease the volume contained within the Fermi surface. Because of defect formation, an imbalance between electrons and holes is expected. Thus, implantation-induced changes in transport properties are expected, both on the basis of band structure effects which would influence the carrier density and effective masses, as well as lattice damage effects which would influence the scattering probability [6.39].

Direct measurements of implantation-induced changes in transport properties have been made by *Yugo* and *Kimura* [6.42], including electrical resistivity, magnetoresistance, Hall effect and thermopower. The transport measurements were made on thin (0.6–1.0 μm) bridge-shaped samples of pyrolytic graphite, implanted at room temperature with C and Ne ions at 40 keV to fluences in the range $3 \times 10^{12} < \phi < 3 \times 10^{16}$ ions/cm^2, using an ion beam current density of about 3μA/cm^2 (R_p = 1100Å and 650 Å, with ΔR_p = 240Å and 280 Å for the C and Ne implantations, respectively). The sample heating during implantation was kept to a minimum by providing good thermal contact between the sample and the substrate. The electrodes were arranged to mainly measure the electrical properties (Fig. 6.21a) of the implanted region, and most of the measurements were made at 77K [6.42]. In this figure, the main rise in resistivity occurs for ion fluences in the range 10^{14} – 10^{15} ions/cm^2.

Several conclusions can be reached from the transport studies. Firstly, ion implantation drastically reduces the carrier mobility. This can be seen from the dramatic decrease in the magnetoresistance with increasing fluence (Fig. 6.21b)

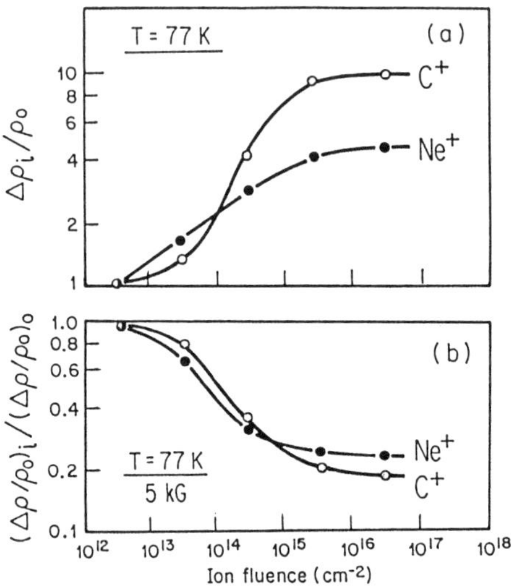

Fig. 6.21. (**a**) Normalized implantation-induced change in the zero field resistivity of carbon- and neon-implanted pyrolytic graphite as a function of ion fluence, where the normalization is to the resistivity of the unirradiated sample. (**b**) Normalized change in the magnetoresistance at 0.5 tesla and 77 K of carbon- and neon-implanted pyrolytic graphite as a function of ion fluence, where the normalization is now to the magnetoresistance for the sample prior to irradiation [6.42]

with large changes taking place near fluences of 10^{14} ions/cm^2. Saturation effects are observed in Fig. 6.21b above a fluence of $\sim 3 \times 10^{15}$/cm^2, suggesting amorphization of the near-surface region, a result that is consistent with other sensitive probes such as Raman scattering [6.4, 30, 32]. Support for the amorphization of the near-surface region was also provided by simulation of the defect density and the defect distribution for the conditions used for the C^+ and Ne^+ implantation, showing that the defect density was equal to the carbon density in graphite (1.1×10^{23}/ cm^3) for fluences of 1.8×10^{15} cm^{-2} for Ne^+ and 2.3×10^{15} cm^{-2} for C^+. The small difference between C^+ and Ne^+ implants, observed in the experiments and also found in the simulation, is attributed to differences in the rate of production of defects per ion for the two cases. Surface cracks also started to appear at an ion fluence of $\sim 3 \times 10^{15}$ cm^{-2} under SEM observations (Fig. 6.9). These cracks are associated with stresses arising from amorphization [6.2, 42].

Results of the Hall effect measurements by *Yugo* and *Kimura* indicate a general shift of the Hall coefficient R_H in the positive direction and a general reduction in the magnitude of R_H, for both C^+ and Ne^+ implantation [6.42]. The Seebeck coefficient determined from thermopower measurements was also observed to be reduced in magnitude and shifted to more positive values [6.42], indicating a higher carrier (hole) concentration. For low fluences, the initial increase in R_H is attributed to a depression of the Fermi levels due to structural degradation, associated with the creation of point defects, and consistent with the magnetoreflection results [6.39]. Although similarities with neutron irradiation studies were found at very low fluences, where the ion implantation is dominated by point defects, major departures were found at high fluences where extended damage and large local stresses occur in the case of massive ion implantation. In comparison, most neutron irradiation studies are carried out in the range where point defect damage persists, because of the low mass of the neutron, its weak interaction with matter, and the higher irradiation energies that are commonly employed.

6.5 Modification of Mechanical Properties

Ion implantation, a technique widely used for improving the mechanical properties of metal surfaces [6.43], has also been applied to increase the surface hardness of graphite, a material with extraordinarily high bulk modulus and high tensile strength in the basal plane but having poor mechanical properties under impact [6.24] (Sect. 5.8). Through radiation damage in the near-surface region, microcrystalline (or amorphous, glassy) material is produced, greatly increasing the surface hardness.

Hardness measurements were carried out by *Hirvonen* et al. on a c-face of HOPG, which was bombarded with 600 keV Xe ions ($R_p \sim 1800$Å) at a rate of 1.9×10^{13} ions/cm^2 s, up to fluences of 10^{14}, 10^{15} and 1.8×10^{16} ions/cm^2. The surface hardness was measured by indentation tests, as the indenter was pushed into the surface at 120 Å /s under various loads until the surface layer fractured [6.44]. The SEM photographs of the indenter tests (Fig. 6.22) show that ion implantation produces a dramatic reduction in the amount of surface damage introduced by the indenter. Deformation of the unirradiated graphite surface takes place by a sliding of the layer planes, relative to one another. But as the near-surface layers become wrinkled and eventually amorphized, this interlayer sliding is inhibited, and the fracture toughness increases. More than an order of magnitude increase in the load-to-fracture is achieved by a fluence of $\sim 10^{15}$ ions/cm^2 (Fig. 6.23) and saturation effects are observed above the onset of amorphization [6.44].

An increase in surface hardness is also found for neutron-irradiated graphite [6.45]. The advantage of using ion implantation rather than neutron irradiation

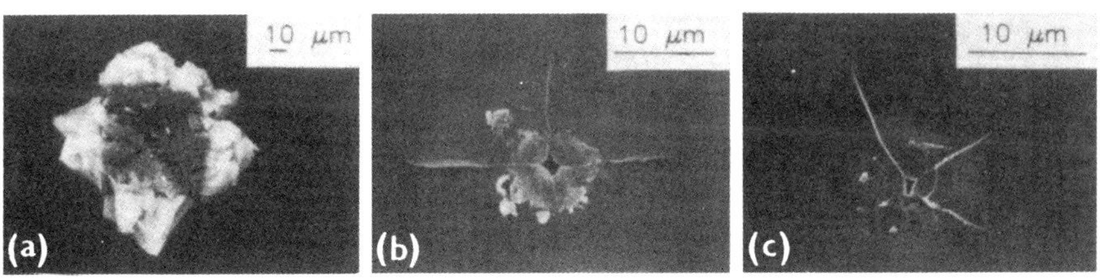

Fig. 6.22. SEM pictures of indentation tests taken by *Hirvonen* et al. on various graphites, (**a**) unimplanted HOPG, (**b**) implanted to 10^{14} ions/cm^2 and (**c**) implanted to 1.8×10^{16} Xe^{++}/cm^2 at 600 keV [6.44]

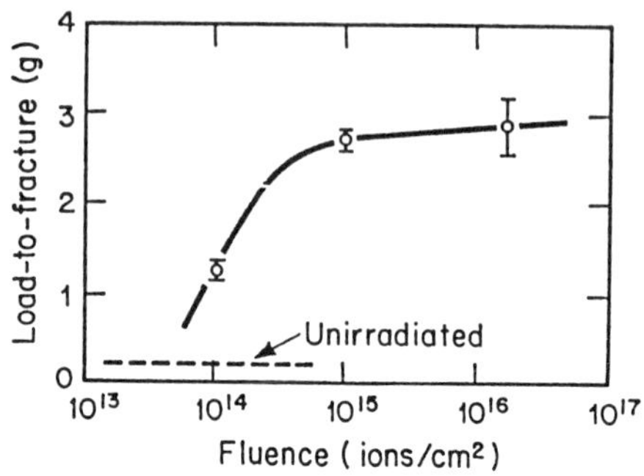

Fig. 6.23. Load-to-fracture as a function of fluence for graphite previously bombarded by *Hirvonen* et al. with Xe^+ ions at 600 keV. The dashed line indicates the load-to-fracture level of graphite prior to ion irradiation [6.44]

is that the amorphous hard layer in the case of ion implantation is very shallow ($\sim$ 2000Å), and the desirable properties of high bulk modulus, tensile strength and high electrical conductivity (in-plane) can be maintained in the bulk, while in the case of neutron irradiation, the neutron penetration depth is substantially deeper and may sometimes even exceed the film thickness.

Measurements by *Matsuhisa* et al. on both PAN and pitch-based carbon fibers showed that ion implantation with 150 keV boron ions to a dose of 10^{15}-10^{17} ions/cm^2 enhanced both the tensile strength and compressional strength of these fibers [6.46]. These improvements in the mechanical properties were explained by Raman microprobe studies on the same fibers which showed that ion implantation greatly reduces the crystallite size, and reduces the crystalline anisotropy, thereby inhibiting shearing [6.46].

Ion implantation of graphite surfaces to high doses eventually leads to surface cracking (Fig. 6.9 and Sect. 6.1) [6.2, 25, 44]. The surface cracking can be explained by the large volume expansion in the region of greatest lattice damage relative to that near the surface, which is less damaged, leading to high surface stresses and eventual surface cracking. Anisotropic swelling (Sect. 6.3) may also contribute to the surface cracking. To enhance the improvement in the mechanical properties of graphite through ion implantation, without the implantation-induced degradation (Sect. 6.1), the microcrystalline (amorphous) structure should be introduced as uniformly as possible.

6.6 Implantation with Ferromagnetic Ions

A thin magnetic film in the near-surface region of graphite can be produced by ion implantation with magnetic species, and the surface can thus be patterned magnetically using a suitable mask. Magnetic films with low coercive forces have been produced by *Koon* et al. by room-temperature implantation of Fe into graphite (HOPG) at 25 keV and at a fluence of 3×10^{16} cm^{-2}. Ferromagnetic behavior was achieved when the Fe concentration exceeded a percolation threshold of $\sim$ 15% [6.47]. Subsequent magnetization and susceptibility studies by *Lusnikov* et al. on samples implanted with Fe at elevated temperature (450°C) and at an energy of 35 keV to a fluence of 2×10^{17} cm^{-2} showed both paramagnetic and ferromagnetic behavior, with the ferromagnetic contribution being relatively insensitive to temperature and the paramagnetic contribution following a Curie law. Analysis of these experiments showed that about 20% of the Fe implant was in the ferromagnetic phase. On the basis of this analysis using SEM characterization of the sample surface, it was concluded that the ferromagnetic phase is associated with small iron clusters ($\sim$ 600 to 800Å in size), which are formed by Fe diffusion to the surface [6.48]. While ion implantation can be used to achieve magnetic behavior in the near-surface region, it is probably not a method of choice for magnetic patterning. The effect of variation of the substrate temperature during implantation needs however further examination to optimize the process.

6.7 Implantation-Enhanced Intercalation

The interest in using ion implantation to enhance intercalation (the insertion of layers of guest species into a host material, see Sect. 2.1.5) arises from the inability to intercalate various species into graphite, as for example the elemental transition metals. On the other hand, since essentially all elemental metallic species can be ion implanted into graphite, it was thought that, perhaps after suitable annealing, the implanted ions would migrate to interlayer positions, causing a lattice expansion along the c-axis, without destroying the in-plane order in the graphene layers. With this lattice expansion along the c-axis, the interlayer coupling between adjacent graphene layers would be reduced, and the implanted ions would subsequently attract similar chemical species provided by the intercalation process.

Exploratory experiments were carried out by *Menjo* et al. to enhance sodium intercalation into graphite, since Na intercalates poorly into graphite (HOPG) and only high stage ($\sim$ stage 6–8) graphite intercalation compounds can be produced [6.49]. By masking approximately half of the sample, and implanting the other half, a controlled experiment was carried out. Indeed some of the implanted ions were found to migrate into interstitial sites between the graphene layers in regions that were not subjected to lattice damage. Although some enhancement to the intercalation process was observed for the introduction of Na into graphite, it was not possible to prepare low stage intercalation compounds in this way. However the enhanced intercalation was found to extend deeper in the sample than R_p. It was furthermore found that similar enhancement in the intercalation of Na into graphite could be achieved by implantation with species other than Na, so that the reduction in the interplanar forces, or the introduction of lattice damage, appears to be more important than the specific implanted species that was introduced or the chemical effects associated with the implant. Though some enhancement to the intercalation was achieved by ion implantation, it was not great enough to merit serious consideration as a practical method for the intercalation of novel species into graphite.

6.8 Implantation with Hydrogen and Deuterium

The effect of hydrogen (H^+) and deuterium (D^+) implantation into graphite deserves special attention for both scientific and technological reasons, which are in part interrelated. From a scientific standpoint, hydrogen and deuterium implantation has a number of unique features. The implantation is carried out at exceptionally low energies, the projectile mass $M_1 = 1, 2$ is much lower than the mass of the target $M_2 = 12$, the implantation doses tend to be exceptionally high, and the implantation temperatures are sometimes very high (exceeding 1000°C). Under typical implantation conditions, some of the implanted hydrogen is transformed into molecular form (e.g., H_2 and CH_4), and these molecules segregate to form gas bubbles [6.50].

Another unique aspect of H^+ and D^+ implantation into graphite is that the fundamental studies are often driven by the use of carbon as the first wall material in fusion reactors (such as tokamaks), where the carbon will be exposed to high doses of implanted hydrogen isotopes. There is thus great interest in understanding the interactions between the hydrogen implants and the carbon targets, especially those interactions that would lead to an erosion of the first wall material [6.51] and to the emission of hydrogen from the near-surface region of the first wall, as well as emission deeper into the wall. As a result of this technological interest, the deposition of energetic hydrogen isotopes into graphite and related materials has been extensively studied [6.52-59].

The special technological issues which must be considered for fusion first wall applications have been extensively reviewed [6.52-54, 60, 61]. These issues include the determination of the amount of the various hydrogen isotopes that are deposited in the wall and their permeation through the wall thickness, especially the possible release of the radioactive tritium isotope. The rate of thermal or ion-induced release of hydrogen isotopes from the surface determines their recycling into the plasma of the tokamak. The outgasing of the walls following a tokamak discharge is especially important in determining the wall conditions encountered by the plasma in the next discharge. Because of these considerations, hydrogen isotope implantation studies into carbon-based materials cover a wide energy range ($10^2 < E < 10^5$ eV) and a wide range of irradiation temperatures ($T_i \leq 2500$ K).

Since $M_1 \ll M_2$ for H or D implantation into C-based materials, the kinematics of the energy cascade is very different from that for heavy ion implants. The energy deposition of the H^+ and D^+ ions takes place near the surface, and for the low energies that are typically used, the nuclear stopping power is significant (for example, for 3 keV D^+, the deposited energy going into nuclear processes is about 20% of the total, with 80% going into electronic processes

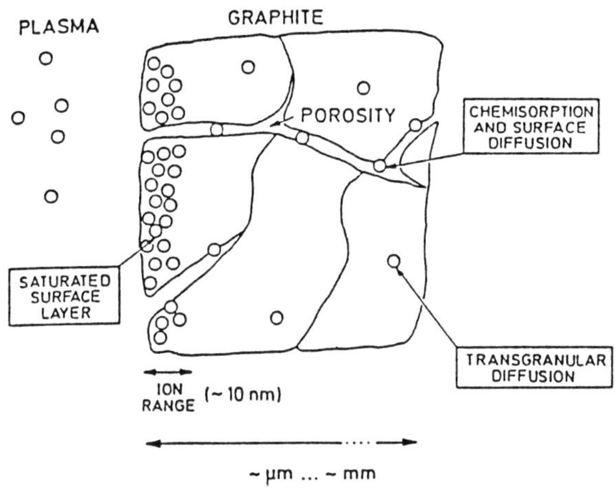

Fig. 6.24. Schematic diagram by *Wilson* and *Hsu* illustrating the principal mechanisms that determine the hydrogen distribution in very high dose hydrogen-implanted graphite [6.53]

[6.60]). From a materials-science standpoint, hydrogen and deuterium implantation present a new set of materials-related issues that are unique to these implants, including bubble formation, molecular diffusion, gaseous emission, and hydrogen embrittlement of the host material. Because of the unique implantation conditions and the special materials-related issues that are encountered, the typical techniques employed for the characterization of hydrogen- and deuterium-implanted graphite are also quite special: thermal desorption, secondary mass spectrometry (SIMS) (Sect. 4.6) and nuclear reaction analysis (Sect. 4.2). For the simultaneous detection of hydrogen and deuterium areal densities, elastic recoil detection (ERD) analysis has been used by several groups [6.62, 63] (Sect. 4.5).

The mechanisms governing the hydrogen retention and distribution are highly temperature-dependent. At temperatures below ~700 K, the hydrogen bombardment results in the formation of a saturated surface layer within the range of the implanted ions, as shown in Fig. 6.24. At somewhat higher temperatures (up to ~1000 K), trapping and diffusion along grain boundaries and pores become important in determining the hydrogen distribution (or inventory), whereas at yet higher temperatures (≥1000 K), hydrogen trapping and diffusion within the grains also takes place (Fig. 6.24). In the saturated surface layer, the hydrogen concentration (which is typically of magnitude H:C ~ 0.4–0.5 at room temperature) exceeds the hydrogen solubility limit and in some cases even exceeds the H concentration in the stoichiometric hydride. By the time the saturation limit is reached, the damaged graphite has been transformed into small grains within a matrix of microchannel networks arising from hydrogen bubble accumulation and diffusion (Fig. 6.24). The saturation effect is illustrated in Fig. 6.25 for deuterium implantation into carbon at room tem-

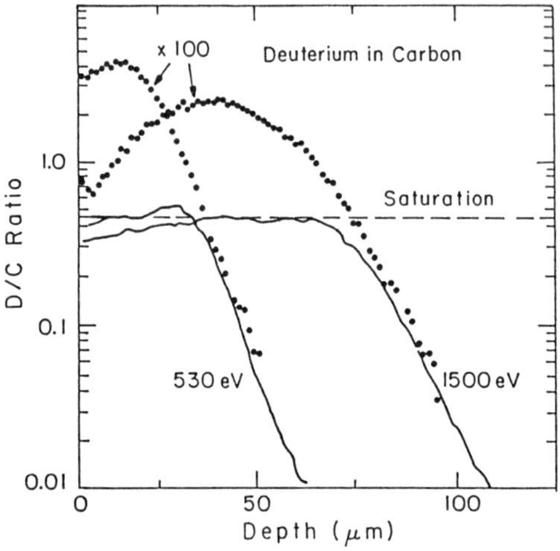

Fig. 6.25. Illustration of the saturated layer effect in depth profiles measured by *Wampler* et al. using the SIMS technique for deuterium implanted into carbon at room temperature for the two indicated implantation energies to a fluence of $10^{18}/cm^2$ (*solid lines*). For comparison, the dotted curves show the profile of $10^{16}/cm^2$ deuterium implanted into carbon where the scale is multiplied by a factor of 10^2 [6.55]. Thus the observation of a saturated layer requires a hydrogen dose far in excess of $10^{16}/cm^2$

perature for a fluence of $10^{18}/cm^2$, where it is seen that the saturation level for the hydrogen detention is independent of the implantation energy, which only determines the depth to which saturation occurs. In this figure the saturation level is monitored by the deuterium to carbon ratio which is measured by SIMS depth profiles (Sect. 4.6).

The saturation effect has been explained by *Möller* and *Roth* using a simple model which limits the bulk concentration to a maximum value above which hydrogen replacement occurs [6.52]. A number of more detailed models have been proposed and applied to explain the experimental hydrogen profiles [6.56]. One of the more promising recent models considers local molecular formation in the implanted layer and subsequent migration of hydrogen molecules [6.60].

From a tokamak applications standpoint, the hydrogen concentration in the saturated layer is found in an all-graphite fusion machine of 100 m^2 wall area to be very high and corresponds to a hydrogen implantation dose of 2×10^{16} H/cm^2 at 100 eV. This dose exceeds by one order of magnitude the surface coverage which corresponds to one monolayer which is $\sim 2 \times 10^{15}$ H/cm^2. The hydrogen in the near-surface of the graphite found in tokamaks is therefore quite unstable and small changes in the hydrogen content of the wall material can strongly influence the behavior of the plasma in the tokamak.

The saturation phenomenon has been studied in an elegant way using different hydrogen isotope implants [6.57-59, 62, 64, 65]. It was found (Fig. 6.26) that the higher the mass of the implant, the larger the detrapping of the hydrogen within the saturation layer [6.62]. Correspondingly, for hydrogen bom-

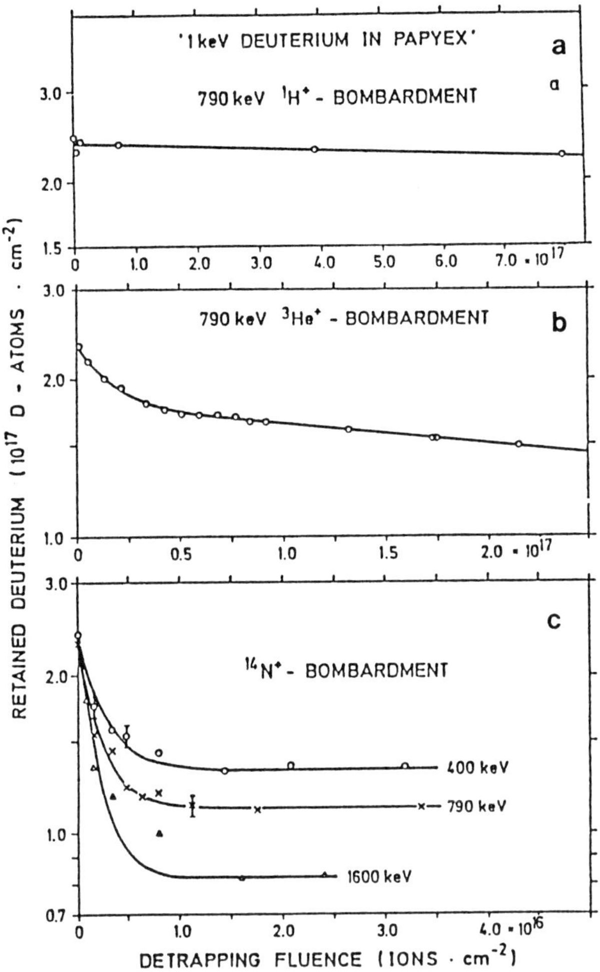

Fig. 6.26. Plot of the retained deuterium atoms (on a log scale) versus the fluence of the hydrogen, helium and nitrogen ions used by *Roth* et al. to study detrapping of previously implanted hydrogen species in carbon. In this figure high energy ions are incident on a nuclear reactor graphite ("Papyex") that had previously been implanted with deuterium at 1 keV to produce a saturated deuterium layer in the near-surface region of the "Papyex" [6.62]

bardment into a nuclear reactor graphite ("Papyex") sample that had been previously implanted with deuterium to saturation, no dramatic effect was observed (Fig. 6.26) at low hydrogen doses in contrast to the effect observed for the heavier ion bombardments. Detailed studies of this hydrogen detrapping effect by *Roth* et al. show that the replacement of H^+ by D^+ in pyrolytic graphite can be modeled by a mixing law, provided that the depth dependence of the distributions are properly taken into account [6.62].

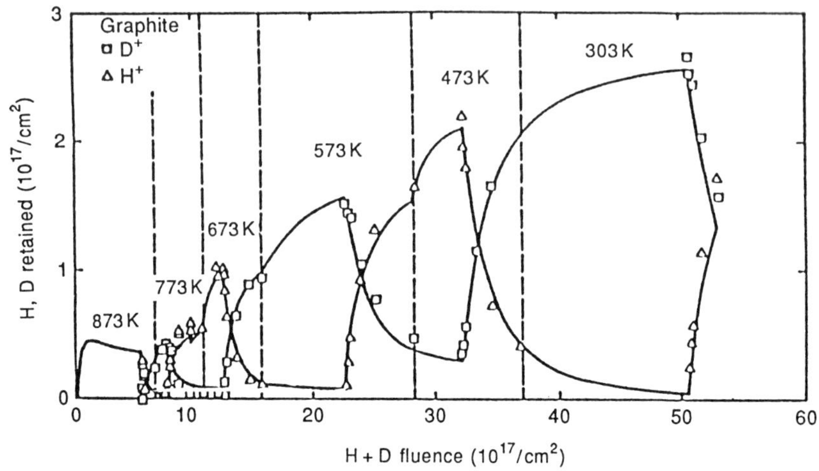

Fig. 6.27. H and D retention in graphite versus implantation fluence during consecutive implantations of H^+ and D^+ at 1.5 keV. The temperature was decreased stepwise as indicated. The buildup of the H concentration and the corresponding decrease in D concentration during H^+ implantation is noted, together with the opposite behavior during the D^+ implantation. The *points* are experimental and the *solid curves* pertain to the good fit obtained using an extended local mixing model [6.66]

Of particular interest to applications is the temperature dependence of the hydrogen isotope retention. To study this effect, consecutive H^+ and D^+ implantations were carried out by *Brice* et al. at 1.5 keV as the temperature was decreased stepwise from 873 K to 303 K [6.66]; the results showed about a 6-fold increase in the H and D retention over this temperature range (Fig. 6.27).

The properties of the implanted saturated layer are in many ways similar to a–C:H layers (Sect. 2.1.8), including optical transparency, low electrical conductivity, and thermal effusion of hydrogen gas at elevated temperatures.

The diffusion of hydrogen molecules through the implanted layer is very fast above room temperature and contributes to the saturation effect illustrated in Fig. 6.25. Gas emission is both in the form of H_2 and CH_4 with the relative methane emission increasing with increasing T. The large accumulation of hydrogen in the grain boundaries leads to hydrogen embrittlement. Although much work has been done in studying the ion implantation of hydrogen isotopes into graphite, the basic mechanisms for hydrogen trapping, de-trapping, recombination and diffusion are still not well understood. In order to describe the hydrogen inventory and the hydrogen recycling in tokamaks, it will be necessary to gain a broad understanding of the implanted layer, the bulk diffusion, as well as codeposition, and chemical and desorption effects [6.60].

7. Implantation-Induced Modifications to Graphite-Related Materials

In this chapter ion implantation studies on various graphite-related (sp^2 bonding) carbons are reviewed. Most of the work on this class of materials has been done on glassy carbon because it is a material of commercial interest that has been well characterized by a number of techniques. Mention is also made of other graphite-related materials that have been studied by ion implantation, though here the literature is sparse except for the carbon-based polymers, which have recently been reviewed in detail [7.1].

7.1 Glassy Carbon

Glassy carbon (GC) is a commercial material that has been relatively well characterized by a number of sensitive techniques. For this reason GC serves as a particularly suitable material to study implantation-induced effects on the structure and properties of a disordered graphite-related material. Ion-implanted glassy carbon is of interest for use in electrodes and in wear-resistant surfaces [7.2].

Prior to ion implantation, GC is described as a porous material (density $\sim$ 1.5 g/cm^2), consisting of a tangle of graphite-like ribbons or microfibrils, $\sim$ 100Å long and $\sim$ 30Å in cross section (Figs. 2.5 and 2.6, where ribbons consisting of somewhat defective fragments of graphene layers are seen). These layers are essentially decoupled in the c-direction (turbostratic stacking) and have high resistivities (4 $\times 10^{-4}$ to 4 $\times$ 10^{-3} Ω cm) relative to graphite [7.3]. Typical heat treatment temperatures for commercial GC materials are T_{HT} $\sim$ 2500°C [7.3]. Because of the tangled network of graphene ribbons, glassy carbon is not a graphitizable carbon.

Detailed studies have shown that ion implantation modifies both the structure and electrical properties of glassy carbon. With regard to structural modification, ion implantation introduces further disorder into the GC, as expected. Once again, the ion mass and ion energy control the profile of the implanted species and the damage profile (distribution) while, for given implantation conditions, the ion dose mainly determines the amount of lattice damage. By choosing different ion species (H, He, C, N, Si, Xe) and implantation energies so that the ion beam modification would peak near 2000 Å in all cases [7.4], it was clearly shown by *Prawer* et al. that the structural damage is closely linked to the number of displacements per host atom (dpa) and not to the chemical

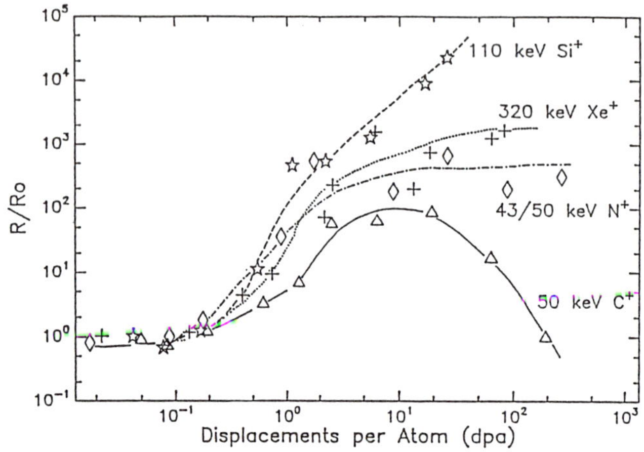

Fig. 7.1. Log-log plot of the resistance (R) of ion implanted glassy carbon normalized to its unirradiated value R_0 as a function of displacements per atom (dpa) for various ions implanted at the indicated energies. The lines are guides to the eye [7.5]

species that is implanted. In other cases, such as the enhanced wear resistance provided by ion implantation to be discussed later in this section, significant chemical effects have also been reported by different groups [7.5, 6].

Surface resistance measurements on glassy carbon by *McCulloch* et al. [7.5] show that the effect of ion implantation is qualitatively different from that on crystalline graphite, insofar as ion implantation changes the resistance of glassy carbon by several orders of magnitude (Fig. 7.1), as compared to the case of graphite, for which the ion-induced change in resistivity is only one order of magnitude. Thus, the resistivity of graphite after implantation is roughly comparable to that of GC prior to implantation. Referring to Fig. 7.1, it is seen that the surface resistance is not sensitive to implantation until the displacement per carbon host atom (dpa) exceeds ~ 0.2 dpa.

The most systematic studies of the implantation-induced damage have been done by *Prawer* et al. with respect to L_a, the in-plane crystallite size. This quantity is sensitively monitored by Raman scattering through measurement of the intensity ratio of the disorder-induced to the Raman-allowed lines at $\sim$ 1360 cm^{-1} and 1582 cm^{-1}, respectively, I_D/I_G, as shown in Fig. 7.2 (Sect. 5.1). For the unimplanted glassy carbon, $L_a \sim 35$Å, while after an implantation dose of $\phi = 3.9 \times 10^{14}$ carbon ions/cm^2 at 50 keV, L_a is reduced to 25 Å. The in-plane crystallite size continues to decrease until, for a fluence of 2×10^{16} carbon ions/cm^2, the Raman spectrum resembles that of so-called amorphous carbon [7.7, 8]. Detailed analysis of the relative intensities of the D and G lines becomes difficult as the ion dose increases because of the dramatic broadening of the lines (Fig. 7.2), though it is clear that some near-neighbor correlations remain because a Raman line (though broad) is seen up to highest fluences in Fig. 7.2. A similar sequence of spectra has also been reported for glassy carbon implanted with Xe ions at 320 keV up to a dose of 9×10^{15} ions/cm^2 [7.4].

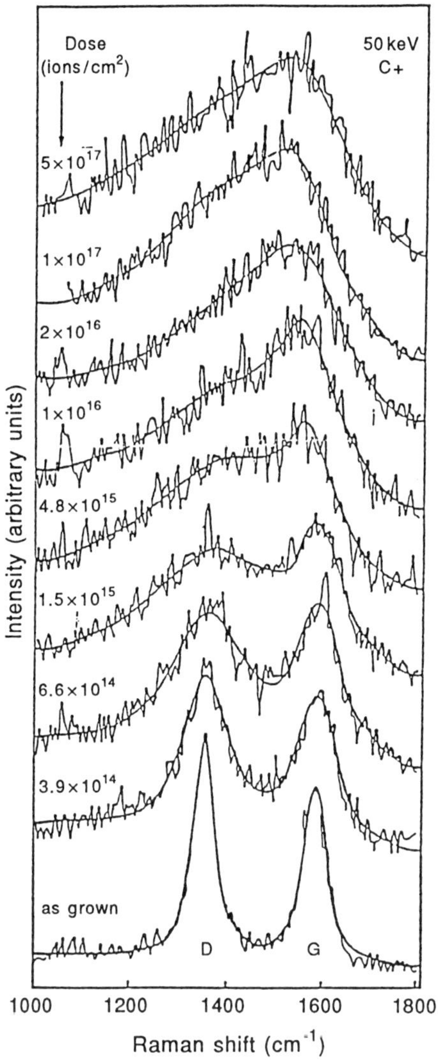

Fig. 7.2. Raman spectra of glassy carbon samples irradiated with increasing doses of 50 keV C^+ ions. The data have been fitted (solid curves) using a two Lorentzian model and a least squares fit algorithm [7.7]

Recent work at yet higher fluences (e.g., $\phi = 1.5 \times 10^{18}/\mathrm{cm}^2$ of C^+ ions at 50 keV) show the reappearance of a double-peaked structure in the Raman spectrum, indicative of an ion beam annealing effect, causing some recrystallization of the amorphized near-surface carbons [7.5]. This interpretation is supported by the drop in resistivity observed above 20 dpa (Fig. 7.1) and the emergence of graphite rings in the RHEED (reflection high energy electron diffraction) pattern for $\phi = 1.5 \times 10^{17}/\mathrm{cm}^2$ carbon ions at 50 keV. This is in contrast to the

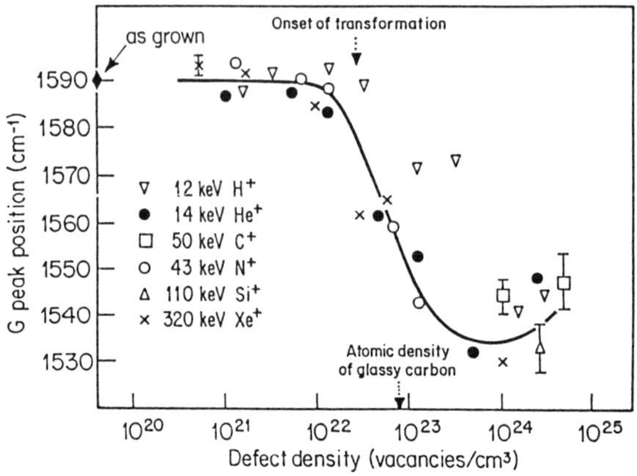

Fig. 7.3. Frequency of the peak position of the Raman-allowed G line for glassy carbon as a function of the logarithm of the implantation-induced defect density for various ions implanted at room temperature at the indicated energies [7.4]

case of 5×10^{16}C$^+$/ cm^2 where broad diffraction rings are observed [7.5], as discussed below. One possible interpretation of this effect is that the amorphized carbon at the surface is more graphitizable than glassy carbon, and ion beam annealing provides enough thermal energy to initiate some graphitic growth. Alternatively, the very high concentration of implanted C ions may be responsible for the build-up of an internal graphite layer centered around R_p, and leading to the observed RHEED pattern.

For doses exceeding 5×10^{17} C$^+$/cm^2 at 50 keV, the temperature dependence of the conductivity displays a maximum at ~250 K, qualitatively similar to that observed in some polycrystalline and pyrolytic graphites. This result confirms the presence of polycrystalline graphite on the surface. The drop in the resistivity, the emergence of RHEED polycrystalline rings, and the change in the temperature dependence of the conductivity for C$^+$ implants were not observed for high dose N$^+$ implants. Therefore the graphitization process is unlikely to be exclusively damage driven.

A detailed analysis of the Raman spectra shown in Fig. 7.2 has been carried out by *Prawer* et al. by fitting each of the spectra to two Lorentzian lines (a D line and a G line), yielding the central frequencies and linewidths for each line [7.4, 7]. This analysis shows that ω_0, the central frequency of the Raman-allowed line, remains essentially unchanged at 1590 cm^{-1} until an ion dose corresponding to 0.21 dpa is reached (Fig. 7.3). At this dpa point, ω_0 starts to fall, showing an approximately linear dependence on the logarithm of the defect density. This linear decrease in ω_0 continues until a value of $\omega_0 \sim 1530$ cm^{-1} is reached, when the displacement per atom is about unity, or equivalently the defect density becomes equal to the atomic density of the carbon host material.

This result can be understood in the following way. At a displacement per atom (dpa) of 0.21, the overall integrity of a carbon hexagon in a graphene ribbon is lost with the introduction of one vacancy per hexagon on the average, neglecting any vacancy-interstitial recombination. This results in an increase in the average C–C distance, and in a corresponding decrease in the C–C force constant leading to a decrease in the vibrational frequency. It is interesting to note that at a dpa of $\sim$ 0.2, the surface resistance also starts to increase dramatically, consistent with the beginning of the destruction of the carbon honeycomb structure [7.5]. At a dpa of unity, the graphitic in-plane honeycomb structure is effectively lost, though the average C–C distance does not seem to change further with yet higher ion implantation doses. On the other hand, the surface resistance is further increased for damage levels beyond a dpa of unity [7.5]. The somewhat high value of the frequency of the Raman-allowed line in unirradiated and lightly implanted glassy carbon is due to the small crystalline size, so that an extended range of k-vectors (near the zone center) contributes to the Raman line intensity, thereby causing an upshift in the frequency of the G-line in accordance with the phonon dispersion relations for single crystal graphite [7.9, 10], as discussed in Sect. 5.1.

Consistent with the Raman studies described above are RHEED patterns, which show (002), (100) and (110) rings in the unimplanted GC samples, but diffuse, broad rings after ion implantation with 1×10^{17} carbon ions/cm^2 at 50 keV. These broad rings bear some resemblance to the spectra seen for amorphous carbon films [7.7]. No detailed, systematic RHEED studies on ion-implanted GC have so far been carried out. One of the reasons for this may be the fact that the RHEED method is not as conducive to quantitative analysis as are the Raman scattering studies, described above.

In keeping with the Raman and RHEED measurements, it would be expected that high resolution transmission electron microscopy (HRTEM) studies of the lattice fringes in GC, a technique which is most sensitive to L_c, would show a monotonic reduction in L_c with increasing ion fluence. This, however, has not been found. HRTEM measurements of the lattice fringes of GC give L_c values in the range 20–40 Å, with relatively little change observed in the TEM pictures upon ion implantation of the GC [7.7] with moderate doses of He^+ ions or large doses of C^+ ions. These results indicate that the basic tangled layered microstructure remains intact (in either the *Jenkins* and *Kawamura* [7.11] form or the *Shiraishi* [7.12, 13] form, see Sect. 2.1.4), even though the in-plane site ordering is largely lost, as is indicated by the Raman results. Supporting this model are radial distribution function data for GC, which show strong evidence for a similarity to crystalline graphite out to the 6 nearest neighbor pairs [7.14]. The HRTEM results may not be completely reliable because of difficulties in preparing suitable thin specimens for HRTEM studies without introducing modifications to the structure of the GC; the HRTEM glassy carbon specimens in [7.7] were prepared by an ultramicrotome technique. Another interesting effect is the partial graphitization of GC by ion beam thinning, as is for example observed when using 5 keV argon ions for sputtering away the GC surface.

With regard to modification of the electrical properties of GC, the effect of ion implantation on the normalized resistance of the near surface region is shown in Fig. 7.1 [7.5]. The very large rise in the surface resistance of glassy carbon (5 orders of magnitude) is qualitatively different from the behavior of

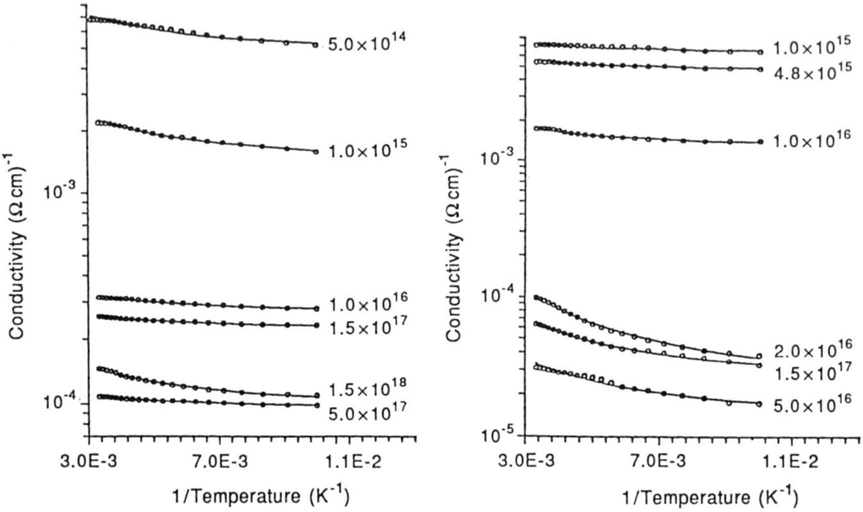

Fig. 7.4. Temperature dependences of the electrical conductivity for GC irradiated with (**a**) 43-50 keV N^+ and (**b**) 50 keV C^+ ions to various fluences. The solid lines are fits to the data using (7.1) for the 3D variable range hopping model [7.16]

ion-implanted crystalline graphite, which is transformed into amorphous carbon at high fluences (Sect. 6.4). Since the resistivity of glassy carbon prior to implantation (Sect. 2.2.2) is comparable to that of ion-implanted graphite or less than that of amorphous carbon, this qualitatively different behavior suggests that ion implantation significantly decreases the amount of graphitization produced during the heat treatment step of the glassy carbon formation process itself. This increase in resistivity in GC following implantation sets in at damage levels above 0.2 dpa, where the integrity of the honeycomb structure of the graphene ribbons begins to break down [7.5]. This experimental observation of an increase in the normalized resistivity can be understood in terms of the calculations of *Bar-Yam* and *Moustakas* [7.15] who show that a high density of vacancies stabilizes the sp^3 bonding.

Further insight into the conduction mechanism is provided by measurements by *McCulloch* et al. of the temperature dependence of the electrical conductivity $\sigma(T)$ of ion-implanted glassy carbon. Such measurements have been carried out using 50 keV C^+ ions (1×10^{15} to $1.5 \times 10^{18}/\mathrm{cm}^2$), 43 keV N^+ ions (5×10^{14} to $1 \times 10^{15}/\mathrm{cm}^2$) and 50 keV N^+ ions (5×10^{16} to $1.5 \times 10^{18}/\mathrm{cm}^2$) [7.16]. The $\sigma(T)$ data for the GC samples implanted up to 1.5×10^{17} C^+/cm^2 were well fit by a 3D variable range hopping model

$$\sigma(T) = \sigma_0 \exp[-(T_0/T)^{1/4}] \quad (7.1)$$

with $T_0 = 8.4 \times 10^4$ K (Fig. 7.4), consistent with $\sigma(T)$ measurements on unirradiated GC (where $T_0 = 4.5 \times 10^4$ K) [7.3], and also consistent with similar measurements on other highly disordered carbons. It is significant that the functional form of $\sigma(T)$ remains the same for ion implanted GC over such a large range of implantation doses. The change in the functional form of $\sigma(T)$ from 3D variable range hopping for doses up to 1.5×10^{17} C^+/cm^2 to metallic conduction for doses above 5.0×10^{17} C^+/cm^2 further supports the formation of a metallic surface layer for high dose C^+ implants into glassy carbon [7.16].

Other anomalies, also observed in Fig. 7.1, are associated with the dependence of the implantation-induced change in surface resistance on the ion species [7.5]. For the case of C^+ implantation, the drop in R/R_0 for lattice damage corresponding to more than $\sim$ 20dpa can perhaps be explained by the regrowth of graphite from the amorphized carbon in the near surface region, or alternatively the build up of a conductive carbon layer due to the high dose C^+ implant. As mentioned above, lattice damage may produce a carbon material near the surface that is more graphitizable than glassy carbon. The saturation effect observed above $\sim$5 dpa for the N^+ and Xe^+ implants may be associated with the complete breakdown of the honeycomb structure of the graphene ribbons, while the increase in R/R_0 for the Si^+ implant up to the highest dpa values may have to do with the formation of some SiC, which is a wide gap semiconductor [7.5]. Thus distinct chemical effects are observed in the transport properties of ion-implanted glassy carbon.

The effect of ion implantation on the mechanical properties of glassy carbon has also been studied by several groups [7.2, 6, 17-19]. It is found that the wear resistance of GC is improved dramatically by ion implantation, as is also the case for graphite [7.20]. Measurement of the surface hardness or wear resistance of ion implanted graphite or glassy carbon samples is difficult because of the thinness of the modified layer and because of the exceptional elastic recovery ($\sim$ 90%) of the substrate. The observation of scratches in the unirradiated material that terminate at the interface between the irradiated and unirradiated regions indicates that the hardness of the ion-implanted material is comparable to or exceeds that of the diamond or B_4C powders used in mechanical wear tests [7.20]. From an applications standpoint, this enhanced hardness is a desirable property of ion-implanted glassy carbon.

Wear resistance measurements for various ions implanted into GC at energies selected to produce a near-surface modification within 1000 Å of the surface have been carried out [7.2], showing that a saturation in the wear resistance is achieved at a displacement energy of 31 $\pm$ 3 eV per target atom and a displacement per carbon atom of dpa$\sim$ 0.5. These results by *Pollock* et al. indicate that the implantation-induced displacement damage is the controlling mechanism responsible for the improved mechanical properties [7.2]. The atom displacements give rise to vacancies, interstitials and dislocation loops, with a high density of small loops formed by room temperature implantation, while fewer and larger dislocation loops are formed with higher temperature implantation [7.2]. The similarity of the magnitude of the displacement per atom (dpa$\sim$ 0.5) associated with the enhancement of the wear resistance to that for the onset of the change in the central Raman frequency (dpa $\sim$0.2–1.0) confirms the role of structural disorder in enhancing the mechanical properties of glassy carbon.

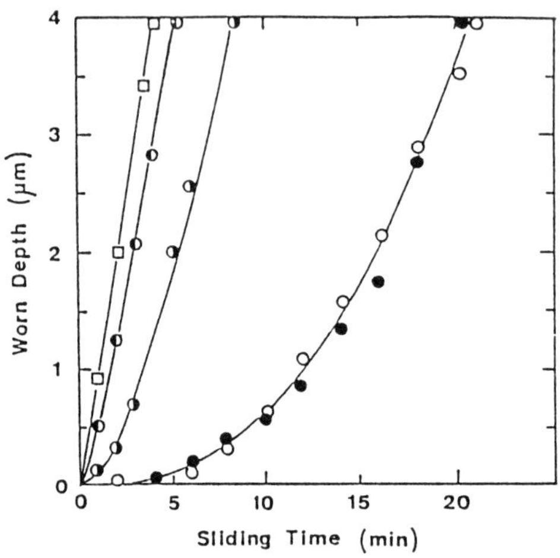

Fig. 7.5. The relationship found by *Iwaki* et al. between worn depth and sliding time for 150 keV nitrogen implanted glassy carbon with fluences of 5×10^{14} ($\oplus$), 1×10^{15} ($\ominus$), 1×10^{16} ($\bigcirc$), 5×10^{16} ($\bullet$) ions/cm^2 at about 25°C. ($\square$) indicates the pristine glassy carbon [7.18]

Iwaki et al. [7.6] have shown the importance of ion species selection for improving the wear resistance of glassy carbon, showing that N^+ implants are highly desirable, though other more reactive species, such as Li, K, F, and Zn are also effective for enhanced wear resistance [7.6]. In particular, the studies of *Iwaki* et al. [7.6] showed that oxygen implantation *reduces* the wear resistance, thereby clearly establishing a chemical effect. By measuring the effect of ion energy on the abrasive wear rate in glassy carbon, *Iwaki* et al. were able to demonstrate the correlation between the increased wear resistance and the increased ion energy (and therefore also with R_p). Similar results were obtained by *Kenny* et al.[7.17]. SIMS analysis of the compositional profile [7.6] showed a migration of the implanted K^+ and Li^+ ions to the surface, thereby explaining why these alkali metal implants were not as effective as the N^+ ions in reducing wear. Amongst the ions considered by *Kenny* et al. (N^+, He^+, C^+), no chemical effect was observed [7.17]. This result, however, does not contradict the findings of *Iwaki* et al.[7.6], who investigated both reactive and non-reactive ion species, and found chemical effects only for the reactive ion species.

A recent study by *Iwaki* et al. [7.18] on the wear resistance of GC using N^+ implanted at 150 keV to doses between 5×10^{14} to 5×10^{16} N^+/cm^2 shows large gains in wear resistance as a function on ion dose, with saturation reached at a dose of 10^{16} N^+/cm^2, as shown in Fig. 7.5. In this figure the material that has been removed from the surface by abrasive action of a 3μm diamond slurry (denoted by worn depth) is plotted against the time for the abrasive action (denoted by sliding time). The authors correlate the onset of the saturation effect with the appearance of the two-peak structure in the Raman spectra (Fig. 7.2), so that the improvement in wear resistance is attributed to the formation of

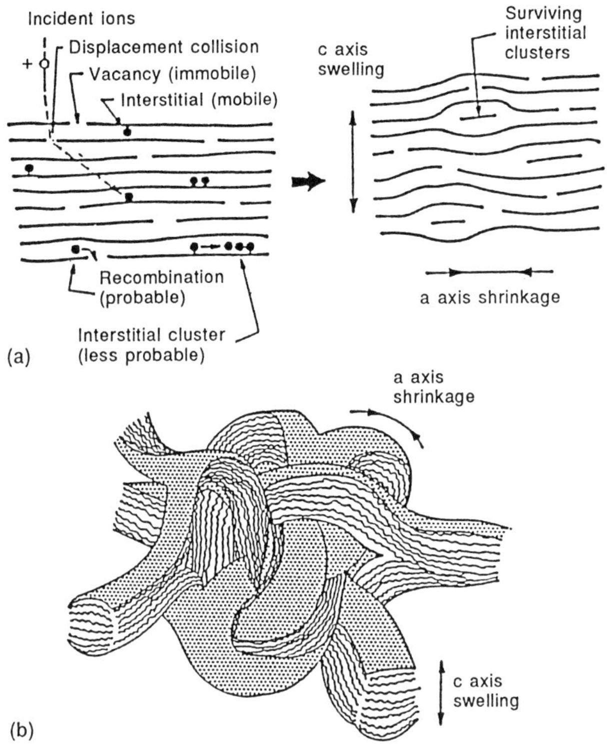

Fig. 7.6a,b. Schematic representation of the effect of ion irradiation on the tangled graphitic structure proposed for polymer-derived carbons according to the model of *Jenkins* and *Kawamura* [7.11]. (**a**) Atomic displacement damage in graphite leading to swelling along the c-axis, shrinkage along the a-axis and hence leading to long-range wrinkling. (**b**) Net result of damage in tangled structure following irradiation. Note, the wrinkling and swelling of planes eliminates voids [7.2]. Though the discussion by *Pollock* et al. [7.2] is based on the *Jenkins* and *Kawamura* model for the glassy carbon structure, similar conclusions about layer wrinkling and void filling are reached on the basis of the *Shiraishi* model (Sect. 2.1.4)

an amorphous carbon surface microstructure. For a given ion dose, greater enhancement of the wear resistance is achieved by carrying out the implantations at reduced temperature (e.g., 200 K).

Several other property changes were found to accompany the enhanced wear resistance, such as increased surface reflectivity, a shrinkage of the size of the implanted region, and a disappearance of the D and G Raman peak structures [7.21, 22]. All of these property changes were found to saturate at the same dpa value (corresponding to $\sim 10^{16}$ N^+/cm^2 at 50 keV) and were attributed to the complete amorphization of the near surface layer by the ion implantation [7.17].

From a microstructural standpoint, the following model for ion implanted glassy carbon emerges [7.2]. Ion implantation results in the creation of immobile vacancies and mobile interstitials (Fig. 7.6a) which can migrate and cluster, leading to local swelling and shrinking, and to long-range wrinkling of the graphene ribbons (Fig. 7.6b). Thus, after ion implantation, GC becomes a tangled network of wrinkled ribbons, which have swelled to occupy most of the original pore volume, with some overall compaction (increase in density) of the implanted region. Sliding between the wrinkled, puckered graphene layers is inhibited by the irregularities, thereby increasing the shear modulus and enhancing the hardness and wear resistance of the implanted surface. In addition, crack initiation is inhibited by the elimination of pores which form weak links in the unirradiated GC. It was suggested that some sp^3 bonds may form between interstitials and carbon edge sites, though this has not been quantitatively established [7.2]. Although this model for the ion-implanted glassy carbon [7.2] formally refers to the *Jenkins* and *Kawamura* model for unirradiated glassy carbon (Fig. 2.5), the conclusions that are summarized above apply equally well to the *Shiraishi* model (Fig. 2.6) delineated in Sect. 2.1.4.

7.2 Carbon Fibers

Carbon fibers have thus far been used for several distinct ion implantation applications [7.23]. However, no systematic study of the implantation-induced modifications of carbon fibers has yet been carried out.

In one application, the special geometry of a vapor-grown (benzene-derived) carbon fiber was exploited to produce highly graphitic TEM samples (Sect. 5.3) where the implantation could be done normal to an a-face (Fig. 6.5), allowing structural studies of L_c to be carried out (Sect. 6.3). Since only the front side of the fiber was implanted, the unimplanted back side of the fiber could be used for calibration purposes for both dark field and lattice fringe images (Fig. 6.5) [7.24].

A related idea was used to measure the effect of ion implantation on the interlayer distance between the graphene layers and the effect of subsequent annealing on this distance [7.25, 26]. These experiments by *Endo* and coworkers showed an expansion of the interlayer distance with implantation that did not anneal out below 1000°C, presumably due to the trapping of the implants at structural lattice defect sites.

Carbon fibers provide an advantageous geometry for the study of the transport properties in graphite-based materials. This advantageous geometry was used to explore the effect of ion implantation and subsequent annealing on the transport properties. Vapor-grown carbon fibers (Sect. 2.1.3) of varying degrees of crystalline perfection (obtained by heat treatment in the range 1800–3500°C) were implanted with boron and phosphorus ions at 150 keV to a dose of $10^{16} - 10^{17}$ cm^{-2} [7.27]. Measurements of the electrical resistivity, magnetoresistance and thermopower were made at 77 K and 300 K for various fibers covering $1800° < T_{HT} < 3500°$C. The temperature dependence of the thermopower was also investigated in some detail. From these measurements it was concluded by *Yugo* et al. that the boron ions act as acceptors and the phosphorus ions act as donors in the implanted carbon fibers. It was also concluded from the annealing studies (carried out at 700°C and 1000°C) that the defects introduced by ion implantation are probably of the same type as those introduced by electron and neutron irradiation [7.27].

From a more practical standpoint, ion implantation has been considered as a means for modifying the near-surface regions of carbon fibers, and hence enhancing their adhesion to matrix materials in a carbon fiber-matrix composite [7.28]. With regard to mechanical properties, ion implantation of 10^{16} N^+ ions/cm^2 at 50 keV enhances the wear resistance of carbon fibers by a factor of 5 for these implantation conditions. However, the enhancement of the wear resistance in carbon fibers is much less than in HOPG, where the wear resistance has been reported by *Kenny* et al. to increase by as much as a factor of 250 [7.17].

Carbon fibers have also provided an excellent vehicle for studying the effect of ion implantation on the tensile strength, compressive strength and torsional modulus of graphite-related materials. It was found by *Matsuhisa* et al. that all three mechanical properties are enhanced by ion implantation of PAN and pitch-based carbon fibers [7.29] as discussed in Sect. 6.5. By making a cross section of the fiber at an angle to the fiber axis, the depth of the implanted region could be monitored by Raman spectroscopy using a Raman microprobe along the cross sectional face [7.29], thereby confirming the ion distribution obtained by the SIMS technique on the same fiber (Sect. 4.6).

7.3 Disordered Graphite

To date, only a few investigations have been carried out studying the effects of ion implantation on the modifications of disordered graphite. The term disordered graphite is used to denote disordered sp^2- bonded carbons, where the amount of disorder can range all the way from weak disorder to amorphous carbon. In fact, one of the difficulties in comparing work on disordered graphites from one laboratory to another is the incomplete characterization of the microstructure of the disordered graphite samples. This small number of studies on disordered graphite may, to some degree, be related to the wide range of properties displayed by disordered graphite films (e.g., values of the resistivity ρ have been reported in the range $10^{-2} < \rho < 10^2$ Ω cm [7.30]) prior to implantation, depending on the substrate and preparation conditions. Working with very high purity but highly disordered graphite films, *Venkatesan* et al. [7.31] found that implantation with 200 keV Xe^+ ions changes the electrical resistivity of the damaged carbon film in the following way. The resistivity remained approximately constant up to a dose of $\sim 10^{14}$ Xe/cm^2, beyond which it started to increase rapidly up to a dose of $\sim 3 \times 10^{15}$/cm^2, above which saturation in ρ occurred at a resistivity value about 10^2 higher than for the unimplanted film (Fig. 7.7). This result for amorphous carbon films is similar in many ways to the corresponding implantation into glassy carbon (Fig. 7.1). For the implanta-

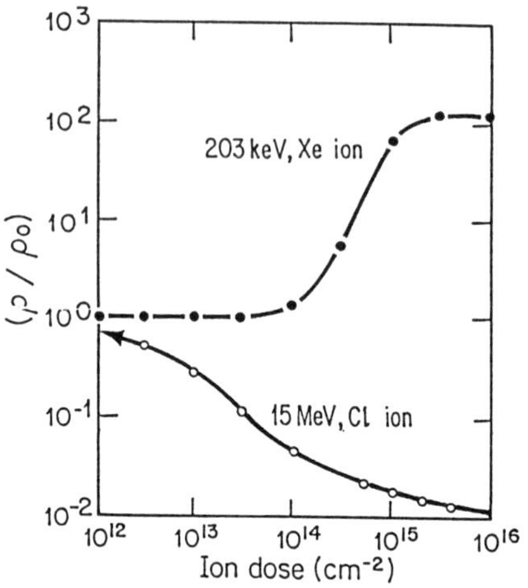

Fig. 7.7. The irradiation-modified resistivity ρ normalized to the pristine film resistivity ρ_0 of an evaporated carbon film subjected to low and high energy ion irradiation, as a function of the ion dose [7.31]

tion conditions in Fig. 7.1, the ion energy loss occurs predominantly via nuclear processes (Sect. 3.1).

However, when high energy implants, for which the dominant energy loss occurs via electronic processes, were incident on amorphous carbon films (such as 2 MeV Ar^+ [7.32] or 15 MeV Cl^+ [7.31]), a decrease in resistivity by 2 orders of magnitude was observed with increasing ion fluence (Fig. 7.7), a highly unexpected observation. Electron energy loss spectra confirm that the high energy ion irradiation enhances the sp^2 bonding. This enhanced electrical conductivity and the corresponding increase in sp^2 bonding are attributed to the huge amount of electronic energy deposited by the ion along its track (diameter ~10Å) [7.31]. *Venkatesan* et al. [7.31] inferred from these findings that microcrystalline graphite had formed along the ion track, thereby providing highly conductive electron paths. Very high densities ($\sim 10^{23}/cm^3$) of localized (very low mobility) carriers were deduced from Hall effect measurements [7.32] on similar samples.

From a more practical vantage point, ion implantation (10^{16} N^+ ions/cm^2 at 50 keV) was shown by *Kenny* et al. to greatly enhance (by a factor of 50) the wear resistance of electrode-carbon which consists of disordered carbon. Although the enhancement measured for implanted disordered carbon was substantially less than for HOPG [7.17], nevertheless such enhancement of the wear resistance could possibly find some practical application since most commercial carbons belong to the disordered carbon family.

7.4 Carbon-Based Polymers

Ion-implantation into carbon-based polymers, in contrast to other graphite-related materials, has been studied extensively [7.1]. Polymers consist of long weakly-bonded carbon chains which are decorated with H, O, N, and other atoms. An important effect of ion implantation on carbon-based polymers is to break bonds and in this way to drive out volatile species, such as hydrogen, oxygen, nitrogen, thereby giving rise to major changes in stoichiometry and in structure [7.1]. With increasing dose, the material shrinks and becomes increasingly carbon rich. Thus, after intense ion-bombardment, the resulting material is similar to a highly disordered carbon, showing electrical properties and Raman spectra similar to highly disordered carbons (Sect. 7.3). For these reasons, there are many common features between ion implantation in polymer-based carbons and in other graphite-related materials reviewed in this chapter. Since ion implantation into polymer-based materials has however been the subject of a recent extensive review article, the reader is referred to this reference [7.1] for a detailed account of this subject. Only a brief summary of this large and important area is provided here.

Early interest in the ion implantation of carbon-based polymers was triggered by the realization that an ion beam could cause either scission of polymer chains or cross-linking between them, thereby either dramatically decreasing or increasing their molecular weights. Corresponding to this change in molecular weight is a change in chemical reactivity which can be utilized for accurate patterning of photoresists (i.e., certain carbon-based polymers) for semiconductor mask applications. For the positive photoresists, implantation enhances the solubility in particular acids, while for negative photoresists, this solubility is reduced. The high spatial resolution now available with current technology with focused ion beams makes ion beam lithography a promising technique for certain sub-micron patterning applications.

The structural and stoichiometric changes introduced by ion implantation cause major changes in the electrical and optical properties of carbon-based polymers. For selected polymers, ion implantation can result in transforming a good insulator into a conducting material, with an increase in the conductivity by more than 10 orders of magnitude upon ion beam irradiation. This is illustrated in Fig. 7.8 by the log-log plot of the conductivity versus ion fluence for ^{84}Br implanted at 200 keV into polyacrylonitrile (PAN, a graphite fiber precursor), and poly-2,6-dimethyl-phenylene-oxide (normally abbreviated by PPO). Thermoelectric power measurements by *Wada* et al. on various implanted polymers show that implantation can yield either p–type or n–type conductors, so that a p–n junction can be made in a polymer through ion implantation [7.34]. The temperature dependence of the conductivity for many implanted polymers is of the form $\sigma = \sigma_0 \exp[-(T_0/T)^{1/2}]$ which is the relation characteristic of the one–dimensional hopping conductivity model for disordered materials.

ESR measurements show that ion implantation introduces an increase in the concentration of unpaired spins as high densities of vacancies and interstitials

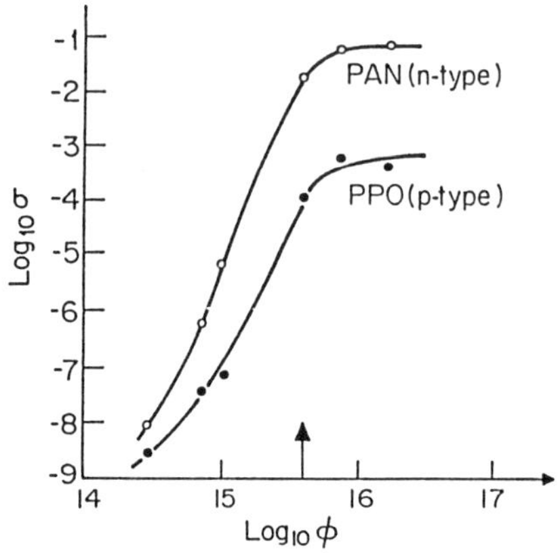

Fig. 7.8. Log-log plot by *Wasserman* et al. of the fluence dependence of the DC conductivity for 200 keV ^{84}Br ion implanted PAN and PPO polymers. The arrow indicates a critical fluence above which the properties do not change much with further irradiation [7.33]

are formed. Plots of the concentration of unpaired spins as a function of ion dose show a qualitatively similar behavior to that shown for the conductivity in Fig. 7.8, including the onset of the increase in spin density and the saturation behavior in the unpaired spin concentration above a critical dose.

With increasing ion fluence, the optical properties of carbon-based polymers also change, as expected from the implantation-induced changes in the transport properties described above; thus, a film that is transparent prior to ion implantation, first becomes cloudy, then darkens to tan, to brown and finally to black at high fluence levels.

From the various measurements briefly mentioned above, it is concluded that energetic ion beams lead to the creation of high densities of broken bonds, free radicals, and localized charges. These implantation-induced charges can carry current by a variable range hopping mechanism. The conductivity becomes significant when the damaged regions of the polymer begin to overlap at the percolation limit. The observation that the conductivity for ion-implanted polymers can exceed that of amorphous carbon indicates the presence of short range sp^2 bonding induced by the ion-implantation-related damage. The similarity between implanted polymers and a–C:H films discussed in more detail in Chap. 9 should be noted.

8. Implantation-Induced Modifications to Diamond

Before going into the details of ion-implantation-induced modifications of diamond, one should emphasize once more the uniqueness of the carbon-based materials, and in particular diamond, as compared to the other systems commonly subjected to ion-implantation. Firstly, carbon is one of the lightest elements ($M_2 = 6$), and thus most implants will be either comparable in mass or heavier, a fact which influences the kinematics of the collision between projectile (M_1) and target atoms (M_2). Hence for the many implantation studies in diamond in which the projectiles were ^{12}C, ^{14}N or ^{11}B, the masses of the projectiles and target are quite similar ($M_1 \sim M_2$), and the conditions for maximum momentum transfer prevail. These may lead to a spatial imbalance between the vacancy and carbon recoil interstitial distributions and this imbalance has been shown to play a major role in the understanding of the lattice damage and its annealing. Secondly, it is unique that diamond, which for many purposes can be regarded as an insulator, may transform upon ion bombardment to graphite, a good conductor. This fact complicates the analysis of implantation effects in diamond insofar as both electrical and optical measurements can yield misleading results, since effects due to the presence of the implant and effects due to graphitization can cause apparently similar modifications to the properties of the starting material. As will be shown, some confusion regarding this point indeed exists in the literature.

The discussion below is divided into the following categories:

1. Basic damage studies in which diamond has been subjected to either carbon or noble gas ion implantation;

2. Damage and annealing studies, mainly for the purpose of utilizing the semiconducting properties of diamond, and obtaining n- or p-type doping, and eventually achieving electronic devices in diamond;

3. Studies aimed at the modification of the optical properties (color) of diamond;

4. Research on implantation-induced effects on the mechanical properties of diamond, utilizing the extreme hardness and high thermal conductivity of diamond and the uses of diamond for cutting edges; and

5. Studies of very high dose implantations into diamond in order to synthesize materials by ion-implantation, especially diamond or SiC.

8.1 Structural Modifications and Damage-Related Electrical Conductivity

Ion implantation into diamond causes macroscopic and microscopic changes which lead to alterations in practically all its physical and chemical properties. The main emphasis of this section is a discussion of the structural modifications which diamond undergoes as a result of implantation; these can in most cases be traced back to the breakage of sp^3 (diamond) bonds and the formation of sp^2 (graphite) bonds. Since graphitization leads to dramatic changes in many of the properties of diamond (e.g., electrical conductivity, density, etc. in Table 2.1), the study of these properties has turned out to be most instrumental in clarifying the picture of the structure of damaged diamond. In the following, therefore, changes in some physical properties of diamond are often mentioned in relation to structural changes. Some of the properties of implanted diamond will be called upon again in other sections under the appropriate headings.

The question of whether and how diamond graphitizes as a result of damage induced by ion-implantation, and what role the implantation plays in this process, have been addressed by numerous researchers. The most commonly employed experimental technique to study the graphitization process in implanted diamonds is to follow the changes in electrical conductivity $\Delta\sigma$ due to increasing implantation dose ϕ, and to correlate $\Delta\sigma(\phi)$ with the formation of graphitic bonds or islands. However, as indicated above, there are pitfalls in this kind of experiment; namely at high implantation doses, conductive layers due to the primary implant can form (i.e., metals or carbon, which is a conductor when in the graphite form). These implants may agglomerate, leading to the formation of buried conductive layers inside the diamond which should not be related to damage-induced graphitization. Indeed, a comparison of the conductivity increases $\Delta\sigma(\phi)$ in diamond and in quartz, which were identically implanted with C ions [8.1], reveal some similarities in $\Delta\sigma(\phi)$ for these two systems, one of which tends to graphitize and the other not. A rough estimate can be made of the dose for the initiation of a buried conductive layer directly caused by the conductive implant. Such estimates show that for light ions, (for example, C^+ implanted at 100 keV, where $R_p \pm \Delta R_p \sim 1370 \pm 220$Å), doses below 1.6×10^{15} cm^{-2} introduce no conducting layers (in agreement with the measurements of C^+ implantation into quartz), while for medium mass ions (for example, Sb implanted at 300 keV for which $R_p \pm \Delta R_p \sim 700 \pm 140$Å), a layer of the conductive species 10Å thick may be formed for doses of the order of 1×10^{15} cm^{-2}. For the heavier mass implants, however, the specified number of broken sp^3 bonds required for the formation of a conductive pathway due to the transformation of the damaged diamond matrix is expected to be reached at substantially lower doses in comparison to light ion implants. This is caused by the higher density of the damage cascade (for example, 870 vacancies/ion for Sb, as compared to 120 vacancies/ion for C implanted at 300 and 100 keV respectively, in Fig. 3.5). Thus for many practical purposes, the measured gross changes in conductivity $\Delta\sigma(\phi)$ can be correlated with structural changes (graphitization) that the dia-

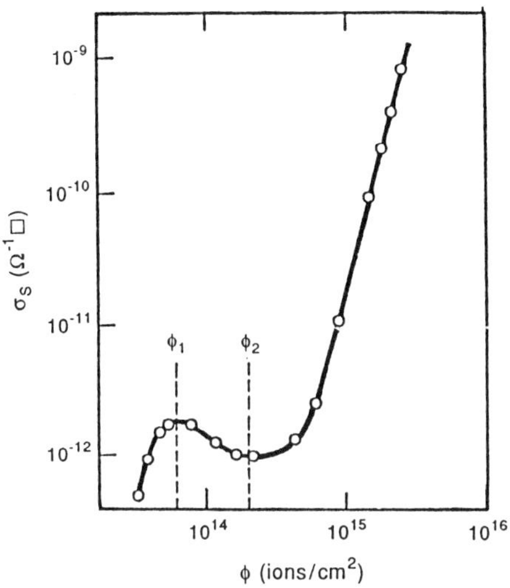

Fig. 8.1. The sheet conductance of the implanted layer of a natural diamond versus 40 keV Ar^+ ion fluence. For $\phi < \phi_1$, there is no formation of an implanted conducting layer below the surface. Partial graphitization occurs in the range between ϕ_1 and ϕ_2, while direct graphitization during implantation occurs above ϕ_2 [8.2] (see text)

mond crystal undergoes which are directly related to the implantation process. The formation of a conductive layer by the implant is of secondary importance only. The "cleanest" experiments in this respect are therefore those in which noble gas ions are implanted, but with care taken to avoid conditions where gas bubbles might be formed inside the diamond due to the limited miscibility of those gases in diamond.

In a pioneering work, *Vavilov* and coworkers [8.2] have correlated implantation-induced changes in electrical conductivity (which they have measured in natural diamond implanted with 40 keV Ar^+ ions at room temperature) with structural changes as inferred from RHEED experiments using 60 keV electrons. The dependence of the sheet conductivity σ_s on Ar^+ dose as given in [8.2] is shown in Fig. 8.1. Three distinct regions are noticeable: for low implantation doses, the conductivity rises to reach a shallow maximum at $\phi_1 \sim 2 \times 10^{14}$ cm^{-2}, it then decreases to a shallow minimum at $\phi_2 \sim 3.7 \times 10^{14}$ cm^{-2}, after which the conductivity rises sharply. This peculiar non-monotonic behavior, first observed by *Vavilov*, has since been found also by other groups [8.3, 4, 5] using other projectiles (Sb, Xe, C) at a variety of implantation conditions. Figure 8.1 thus seems to present a universal behavior of ion-damaged diamond. From their RHEED experiments, *Vavilov* and coworkers deduced information about the structural changes that are related to the observed changes in conductivity. In the low dose regime, in which isolated point defects are speculated to be responsible for the conductivity rise, the RHEED pictures revealed the presence of small damage clusters in the shape of 2–6Å thick disks, buried ~ 100Å below the diamond surface. The rise in conductivity observed in the low dose regime is attributed to the presence of these clusters. With increased implantation dose ($\phi \sim 7 \times 10^{13} - 3 \times 10^{14}$ cm^{-2}), which covers the region where the conductivity drops, partial ion-induced graphitization takes place. The RHEED patterns [8.6] show that here the affected volume seems to consist of small weakly misoriented diamond blocks ($\sim$ 1000Å in size) and twins which break up into smaller blocks with further implantations. For doses around the minimum in conductivity $\phi \sim 3.7 \times 10^{14}$ cm^{-2}, the electron diffraction pattern practically disappears. However, with even higher implantation doses, intense graphitization sets in and the whole volume affected by the implantation ($R_p + \Delta R_p \sim$ 300Å) consists of weakly oriented small graphite crystallites with an average crystallite size of $L_a \sim$ 100Å and $L_c \sim$ 30Å. The formation of these graphite crystallites was taken in that work [8.2] to be responsible for the large increase observed in the electrical conductivity.

The fact that, following high dose implantation, graphite islands are formed in the diamond and that these islands are responsible for the large measured electrical conductivity has been verified by *Hauser* and coworkers [8.7, 8] who have measured the temperature dependence of the electrical conductivity $\sigma(T)$ in heavily carbon-ion-implanted diamond. Natural diamonds have been implanted [8.8] at room temperature with selected doses of C^+ ions at 70, 40 and 20 keV, so as to build a fairly uniform C implant profile in the specimens; the implantation doses ranged from 3×10^{15} to $6 \times 10^{16} C^+/cm^2$. The temperature dependence of the resistivity $\rho(T)$ for these high doses, measured from room temperature down to 20 K, are shown in Fig. 8.2. The data in Fig. 8.2 fit the relationship for variable range hopping conductivity

$$\rho(T) = \rho_0 \exp(T_0/T)^{1/4} \tag{8.1}$$

very well over the entire temperature range. Departures from this variable range 3D hopping model were found only at high temperatures and for the lowest dose (3×10^{15} cm^{-2}), in which case, the temperature dependence $\rho(T)$ was found to follow the relation

$$\ln \sigma(T) \; \alpha \; (E_0/kT) \tag{8.2}$$

typical of a thermally activated process. Here $\rho(T) = 1/\sigma(T)$ and the activation energy deduced from the data is E_0 = 1.7 eV [8.7, 8]. At the highest dose (6×10^{16} cm^{-2}), the density of states deduced on the basis of a 3D variable range hopping conductivity model turned out to be unreasonably high, while the measured resistivity was very low and nearly temperature independent; hence the authors concluded that metallic conductivity governs this regime, suggesting an implantation-induced graphitic layer, at least with regard to its electrical properties. Nevertheless, the hardness of the implanted layer was still close to that of diamond. These facts, i.e., that the implanted region behaves electrically like graphite but mechanically like diamond, were explained by *Hauser* to be due to the relatively high density of graphitic (sp^2) bonds (10^{20} cm^{-3}, a density which is beyond the percolation threshold) that are responsible for the high conductivity. The majority of the bonds, however, are still undisrupted diamond

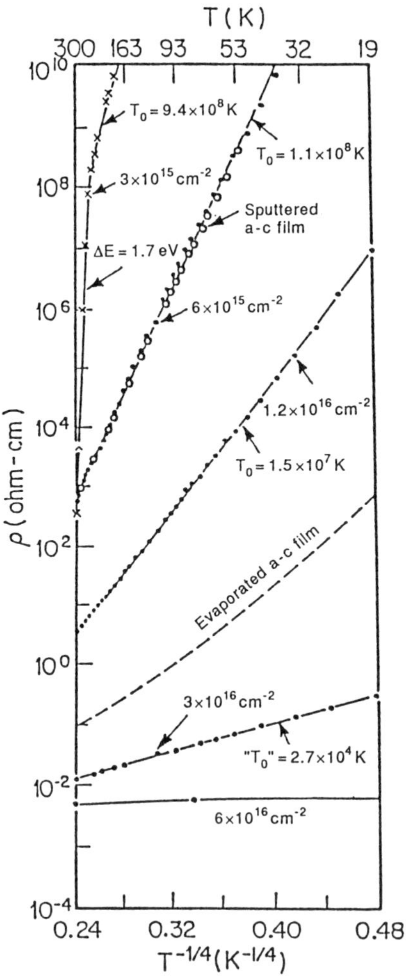

Fig. 8.2. Temperature dependence of the resistivity for various a–C (amorphous carbon) layers. 1100 Å a–C sputtered at 95 K and annealed at 300 K; a–C film evaporated at 300 K; implanted diamond with increasing implantation doses. (The doses listed represent total dosage in C^+/cm^2) [8.8]

bonds (sp^3) and they are responsible for the diamond-like mechanical properties of the implanted region.

A different approach to the onset of conductivity in ion-implanted diamond was proposed by the Technion group [8.5]. In this work the rise in room temperature conductivity with increasing implantation dose was studied for 340 keV and 160 keV Sb ions for different implantation geometries (Fig. 8.3). While the data of *Kalish* et al. [8.5] closely resemble those of *Vavilov* obtained for Ar implantation [8.2], their interpretation was different. The onset of electrical conductivity, which was observed [8.5] to occur at $\phi \sim 10^{14}$ cm^{-2}, was analyzed in

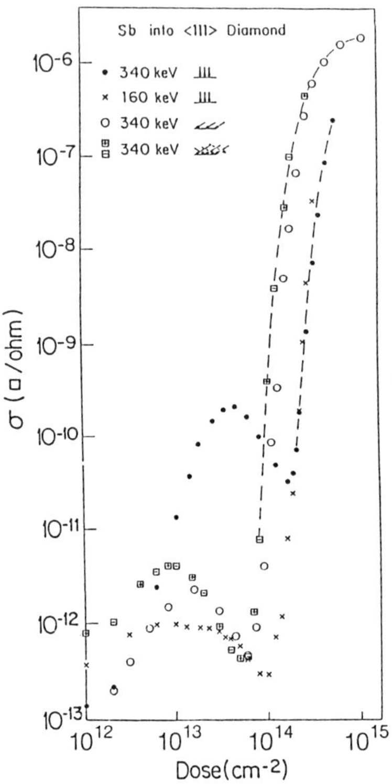

Fig. 8.3. Sheet conductivity of Sb-implanted diamond versus dose for various implantation conditions. The legend in the inset gives the energy of the implant, and the symbol on the right tells whether the implant was done at normal incidence or at some other angle (+ or − 60° with respect to the normal) [8.5]

terms of percolation theory, assuming that electrical conductivity sets in once a percolative pathway (i.e., connectivity between conductive regions) is reached. According to this analysis, variable range hopping conductivity between conductive regions was found to fit the data very well [8.5]. By performing the experiments at different implant energies and at different target tilt angles, the authors were able to clearly show that the important factor in transforming insulating diamond into a conductive form of carbon is the *energy density* in the damage cascade of each ion. Furthermore, it was deduced from the steepness of the percolation transition that several overlapping cascades are needed to turn diamond into a highly conductive form. Basic information on the size and shape of the damaged volume was deduced from the data. The reasons why graphite is not immediately formed during ion impact are proposed [8.5] to be due to the high internal pressure in which the damage cascade is created inside the diamond. This pressure suppresses graphitization. The existence of a threshold density of graphitic bonds in the damaged diamond matrix required for the total collapse into the graphitic phase may also inhibit immediate graphitization.

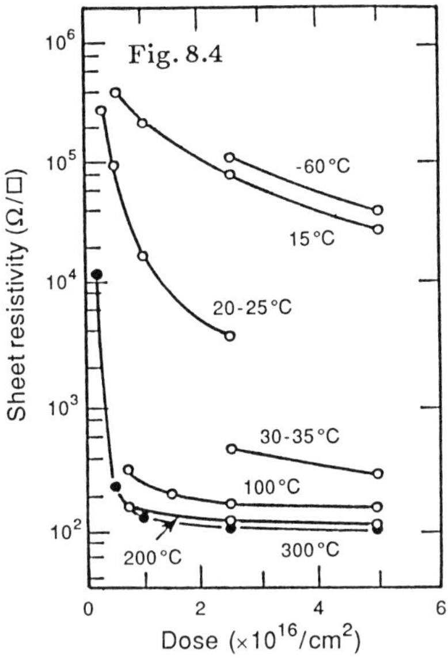

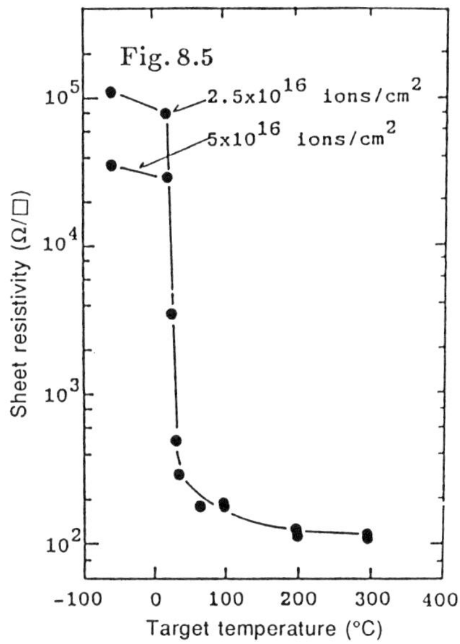

Fig. 8.4. Sheet resistivity of ion-implanted diamond (Ar^+ at 150 keV) as a function of Ar^+ dose at various target temperatures [8.10]

Fig. 8.5. Sheet resistivity as a function of target temperature for ion-implanted diamond (Ar^+ at 150 keV and doses of 2.5×10^{16} and $5 \times 10^{16}/cm^2$) [8.10]

The onset of hopping conductivity in carbon-implanted diamond has been analyzed by *Prins* [8.9], under the assumption that for implantations at elevated temperature (240°C) interstitials may diffuse out of the damaged region, while vacancies are locked in, being immobile at this temperature. Hence, a vacancy-rich region having a thickness of approximately the implant range straggling ΔR_p is formed around R_p. *Prins* assumed that this region collapses at a certain critical dose into a conductive layer where charge carrier hopping occurs. Based on statistical and diffusion considerations, *Prins* concluded that the onset of conductivity should depend on the implant profile (and through it on the implant energy) according to $(\Delta R_p)^{3/2}$, a dependence which he has indeed verified experimentally [8.9].

Elevated implantation temperature (T_i) or high temperature post-implantation annealing (T_a) enhances the onset of conductivity, possibly because of the enhanced formation of graphite in the implanted region. In this connection, *Sato* and coworkers [8.10] observed an extremely sharp dependence of the electrical conductivity of Ar-implanted diamond (150 keV) on target temperature T_i (Figs. 8.4 and 8.5); the dependence of the resistivity on T_i is steepest just around room temperature, the measured resistivity dropping by over two orders of magnitude between 15°C and 35°C for identical implantations ($2.5 \times 10^{16} Ar^+/cm^2$). This high sensitivity to implantation temperature near 300 K makes the quantitative comparison between different experiments carried out at so-called "room temperature" extremely difficult. Slight variations in implantation power (i.e., beam current or energy) or differences in diamond heat sinking may, in the light of this finding, cause a very large spread in the properties of ion-implanted diamond when measured on different samples implanted under the same nominal implantation conditions.

Post-implantation annealing of diamond samples implanted at temperature T_i up to 500°C shows a nearly logarithmic drop in resistivity with measuring temperature T for samples implanted at –60°C or +15°C, but hardly any change in $\rho(T)$ for those implanted at 35°C, 100°C and 200°C. The reason for the observed temperature dependence of $\rho(T)$ may be due to the diffusion of interstitials in diamond, which becomes important above $\sim$ 50°C, or due to the formation of a graphitic material which may happen around room temperature. The onset temperatures for both processes are expected to be lowered by the thermal spike associated with ion implantation.

The results of experiments carried out by *Prins* [8.11] in which the resistivity changes $\Delta\rho(\phi)$ were measured on carbon ion-implanted diamond for two implantation temperatures ($T_i = -196$°C and +240°C) are even more striking than those of *Sato* et al. [8.10]. Above an implantation dose of 10^{15} cm^{-2} for 30 keV C^+ ions, where the resistivities $\rho(\phi)$ for both cold- and hot-implanted diamonds saturate, *Prins* found that the cold-implanted diamond exhibited a sheet resistivity 6 orders of magnitude higher than that for the hot-implanted diamonds. Isochronal annealing of the highly resistive sample at $T_a \sim 400$°C was found to lower its resistivity to that of the hot-implanted diamond. *Prins* concludes that even though both implanted samples look pitch black, and even though both implanted regions must have been totally amorphized by the high density of the carbon implants, the nature of the amorphized layers must be very different for the two samples. In the amorphous layer created at low implantation temperatures T_i, no rearrangement of atoms and bonds is possible; hence it does not show high electrical conductivity nor does it exhibit any temperature dependence $\rho(T)$. This is in contrast to implantation at elevated temperatures where an immediate relaxation of the damage to some conductive form takes place. *Prins* concluded that thermal energy must be supplied to the amorphous layer to assist in the formation of the necessary conductive bonds, which are the basis of the hopping conductivity observed for these specimens.

Thus, from the various studies of the temperature dependence of the conductivity of ion-implanted diamond, important information on the nature of the radiation damage in diamond can be obtained and the conduction mechanisms operative under the various implantation conditions commonly employed in the modification of the properties of diamond by ion implantation can be studied.

8.2 Volume Expansion

So far the relationship between changes in the diamond structure and changes in its electrical conductivity induced by ion implantation have been discussed. The explanation for the increased conductivity offered by most researchers was

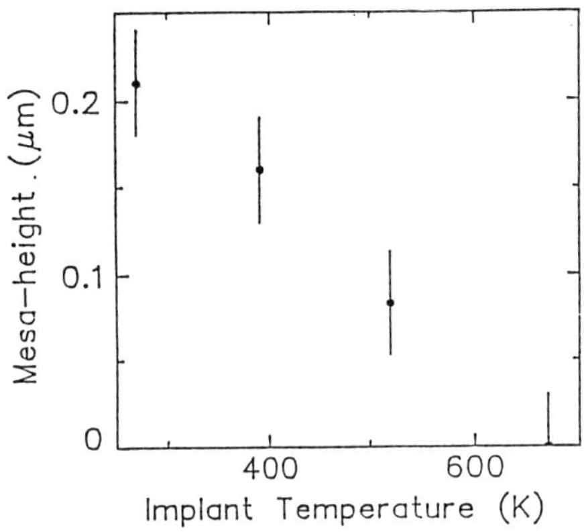

Fig. 8.6. Plot of mesa step heights versus implant temperature T_i for 120 keV C^+ implanted into diamond to a dose $\phi = 5 \times 10^{16}$ cm^{-2} [8.15]

that the formation of sp^2 bonds and their collapse to graphite, which takes place above a certain critical density and which requires the availability of thermal energy, were responsible for the large measured increases in conductivity and for their temperature dependence. However, another phenomenon, which may not just be related to simple graphitization, has been found by *Maby* and coworkers [8.12] and by *Prins* et al. [8.13, 14] to accompany high dose implantations of diamond, namely lattice swelling. This implantation-induced expansion is noticeable as steps between implanted and non-implanted regions in diamond and exhibits a strong dependence on implantation temperature as can be seen in Fig. 8.6 in which step heights versus T_i are shown for 120 keV C^+ ions implanted to a dose of 5×10^{16} cm^{-2} [8.15].

The ion-induced volume expansion measured for the tightly packed diamond is more complicated than that found for implanted graphite, with its lower density. Three factors may, in principle, contribute to increases in volume due to the implantation: the addition of extra atoms, forced into the lattice by the implantation; the creation of vacancies and interstitials in the collision cascade; and the possible phase transformation from compact diamond (density of 3.515 g/cm^3) to the less compact but more thermodynamically stable graphite (density of 2.26 g/cm^3) enhanced by sp^3 bond breakage by the traversing ion.

Maby et al. [8.12] have implanted diamond with boron ions and have reported an irreversible volume expansion which sets in above a certain critical dose ($\phi > 5 \times 10^{15}$ cm^{-2}). They attribute their observation to the creation of an amorphous region due to the ion-induced damage which has the graphite density, thereby accounting for the observed swelling.

A different approach to the processes which determine the physical and structural properties of light ion implanted diamond has been given in a series of recent publications by the South African group [8.13, 14]. The basic point in explaining both the volume expansion and the increase in electrical conductivity measured for ion implanted diamonds is, according to this work, the difference in the spatial distribution of the vacancies and interstitials created in the damage cascades and the difference in their mobilities. While substantial diffusion of self-interstitials already occurs at about 50°C during implantation [8.11], the vacancies formed during the collision cascade are immobile until about 500°C. As a result, diamonds implanted at low temperatures ($T < 50$°C) should show little swelling while those implanted at elevated temperatures (50°C $< T <$ 500°C) should exhibit volume expansion, due to the escape of interstitials from the damage region with the vacancies still locked in. Indeed, the Johannesburg group demonstrated that higher steps are formed for diamonds implanted with 170 keV ^{19}F ions at elevated temperatures [8.13] than at low temperatures (77K), and that ^{13}C diffusion is correlated with this step height [8.16]. At high temperatures ($\sim$ 400°C) no steps could any longer be measured between the implanted and unimplanted regions, and substantial ^{13}C in- and out-diffusion was observed. *Prins* et al. [8.13] also give a series of arguments to explain why graphitization is an unlikely process for explaining the swelling of implanted diamond. Based on the number of graphitic bonds estimated by *Hauser* et al. [8.8] to be responsible for the hopping conductivity measured by them for implanted diamond, only negligible swelling could result even if all those bonds were to form graphite crystallites. The rate of volume expansion measured by *Prins* et al. [8.13] for the case of ^{19}F implanted diamond at $\sim$ 100°C shows a distinct change in slope which sets in at a dose of $\sim 5 \times 10^{16}$ cm^{-2} (Fig. 8.7). Since part of the volume expansion must occur to accommodate the implanted F atoms in the diamond matrix, and this should be approximately linearly related to the ion dose, Fig. 8.7 should actually be composed of the superposition of a linear part due to the introduction of extra F atoms into the diamond and an additional part related to structural changes in the damaged diamond. When this decomposition is done to the data of Fig. 8.7 and a linear contribution is subtracted from the data, a curve exhibiting a sharp rise at low doses, which saturates at a dose of $\sim 5 \times 10^{16}$ cm^{-2}, is obtained (Fig. 8.8). This sharp rise in volume expansion observed for relatively low doses is suggested by *Prins* et al. [8.13] to be due to the out-diffusion of interstitials from the implanted volume. From the slope of the linear part of Fig. 8.7, *Prins* and coworkers [8.13] deduce that the linear expansion is equivalent to a volume increase of 26 Å^3 for each injected ion. (It should be noted that this number is $\sim$ 5 times larger than the atomic volume of each ^{19}F atom!)

Purely theoretical calculations on self-diffusion mechanisms in diamond exist, and have been reviewed in a recent paper by *Bernholc* et al. [8.17]. Interestingly, it is found from this calculation [8.17] that the self-diffusion in diamond is dominated by *vacancies*, which are predicted to be the most mobile species in diamond. This is apparently in contrast with the findings of the South African group [8.9] described above, which are based on studies of the swelling of diamond [8.13] and on tracer experiments on the diffusion of ^{13}C [8.18], both of which can be understood if *self-interstitials* are the most mobile defect in diamond. The difference between these two orthogonal ideas may have to do with the fact that the theory [8.17] deals with an undamaged diamond crystal,

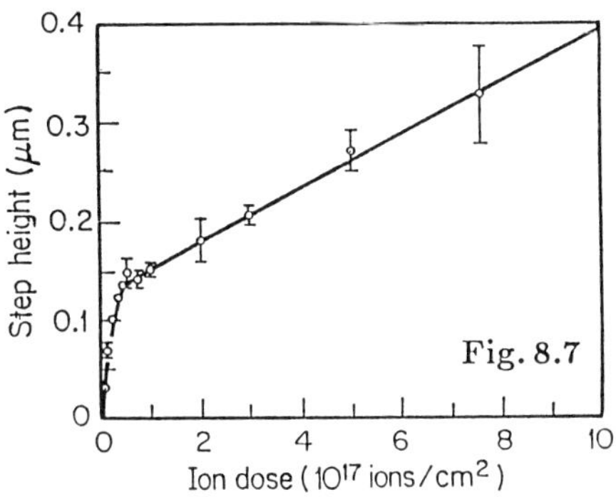

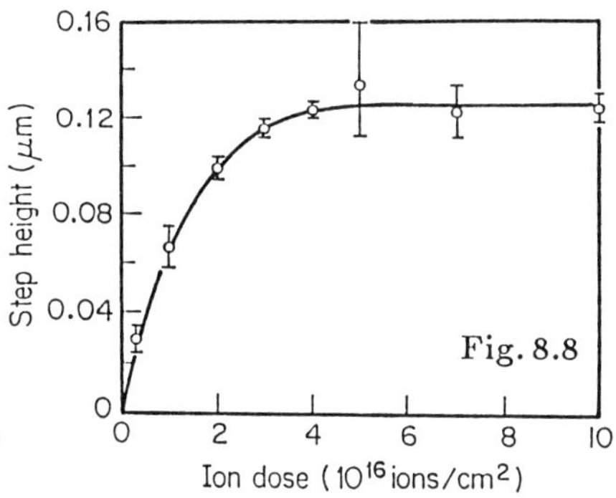

Fig. 8.7. Dose dependence of the step height on a diamond surface caused by the volume expansion during implantation of 170 keV fluorine ions [8.13]

Fig. 8.8. Dose dependence of the rapid volume expansion observed at low ion doses after subtraction of the linear expansion which is prevalent at high ion doses [8.13]

while experiment refers to the dynamics of diffusion under ion bombardment, which corresponds to a severely damaged crystal. Nevertheless, this apparent discrepancy is worth further consideration.

A statistical model, based on the assumption that the volume expansion at low ion doses should be a direct measure of the number of immobile vacancies remaining in the damaged layer after out-diffusion of the interstitials, allowed *Prins* et al. [8.13] to deduce the displacement energy E_d for a C atom in diamond. E_d represents the minimum energy that is needed to dislodge a carbon atom from its lattice site and to move it far enough from its vacancy so that recombination is no longer possible. This parameter, which is an important input parameter in all simulation programs (i.e., TRIM), is found by the above analysis to be $E_d \sim 55$ eV in diamond, in rough agreement with previously published values of 80 eV [8.19] and more recently 35 eV [8.20]. The physical picture of the swelling of the implanted diamond [8.9], namely that following implantation at ambient temperature the implanted ion may reside in a vacancy-rich environment, has important implications for the possibility of annealing the crystal. In order to obtain perfect annealing, as is for example needed for electrical activation of implanted donor or acceptor species, it may be important to supply the vacancy-rich regions with extra carbon atoms to annihilate the vacancies. This point is discussed in more detail in Sect. 8.4.

The spatial distribution of vacancies and interstitials should, in the light of the above arguments, play an important role in the final state which an implanted diamond layer assumes following low or room temperature implantations. Cross-sectional transmission electron microscopy (XTEM) should, in principle, be able to reveal vacancy-rich and interstitial-rich regions in the implantation-affected volume. However, such measurements have not yet been performed on diamond, mainly because of the difficulty in preparing suitable

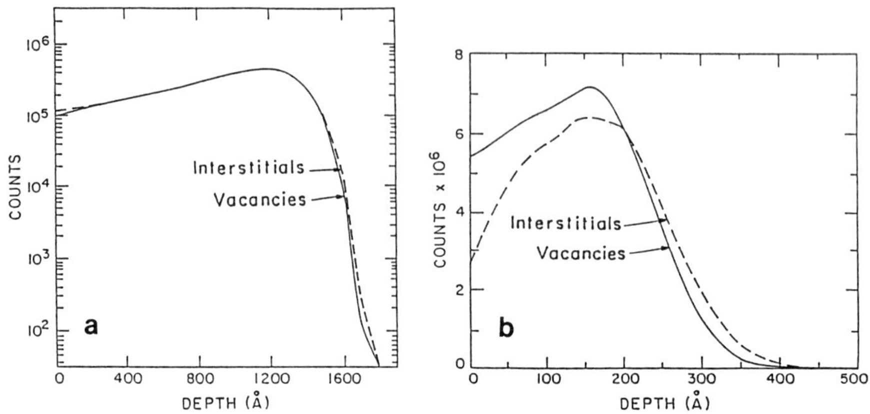

Fig. 8.9. The results of Monte-Carlo simulations of the vacancy (*solid curves*) and interstitial (*dashed curves*) distributions following (**a**) 100 keV ^{12}C and (**b**) 100 keV ^{209}Bi implantations into diamond. The count scales should be multiplied by the implantation doses to obtain the actual distribution (in vacancies or interstitials per cm^{-3}). Note that the counts for the ^{12}C implant are given on a log scale while the counts for the ^{209}Bi implant are on a linear scale

specimens. The final atom distribution following implantation can, however, be simulated by the use of Monte-Carlo simulation computer codes such as TRIM (Sect. 3.4 above). Since TRIM is capable of storing the complete history of each collision cascade, it can also yield the net displacement of host atoms following implantation. TRIM has been modified by *Shaanan* [8.21] to do just that. The program stores all locations at which displacement collisions take place—i.e., the locations at which vacancies are formed. It then follows the recoils until they come to rest at some new, on the average deeper, location. Plots of the vacancy and interstitial distributions can thus be obtained. The differences between two distributions yields some vacancy-rich regions, where the density of the implanted layer is below the normal bulk density, and some regions with densities above normal due to excess interstitials. Figures 8.9a and b show the results of such calculations for the case of 100 keV C ions (Fig. 8.9a) and for 100 keV Bi ions (Fig. 8.9b) implanted into diamond. For these simulations E_d=55 eV was used. The calculations were terminated for ion energies below 1 keV since for lower energies the recoils are not expected to be much separated from their origin, and hence these recoils would not much affect the final net distribution of surplus vacancies or interstitials. The two distributions shown in each figure are those of the vacancies (solid curves) and of the interstitials (dashed curves). It is obvious from Fig. 8.9b that for the case of a heavy projectile (^{209}Bi where $M_1 \gg M_2$), significant vacancy and interstitial-rich regions are predicted by the simulation. For example, a 100 keV Bi implantation to a dose of 10^{14} cm^{-2} should lead to an excess vacancy concentration of about 3×10^{20} cm^{-3} near the surface and to about 1×10^{20} cm^{-3} at a depth of 150 Å, whereas for the case of a light projectile such as ^{12}C ($M_1 = M_2$), the excess vacancy concentration over the first 1000 Å (depth $\leq R_p$) is substantially smaller. Nevertheless, this imbalance may well yield a measurable swelling of the implantation-affected volume.

If the above simulations, which neglect any recombination or diffusion, are correct, and if the swelling of implanted diamond is predominantly due to vacancy-rich regions, then diamond implanted with heavy ions should swell substantially more than when implanted with light ions. This has indeed been verified by *Prawer* and *Kalish* in a recent experiment [8.22] in which diamond was implanted at room temperature with 100 keV ^{12}C and 320 keV ^{133}Xe ions and the step height between the implanted and unimplanted regions was measured. It was found that a swelling of the diamond by about 900 Å was induced by the implantation of 10^{17} C ions/cm^2 while a dose ten times less (10^{16} ions/cm^2) was sufficient to obtain the same swelling following Xe implantation.

8.3 Lattice Damage

Direct information on the kind of damage in diamond and on its build-up with increasing implantation dose can be deduced from light ion (H or He) channeling experiments. Since crystalline diamond has a simple structure and a very high Debye temperature (i.e., low amplitude thermal vibrations), the channeling effect in diamond is very pronounced, and extremely good channels (low values of $\chi_{\min}$) can be found in diamond. Particularly nice channeling patterns are exhibited by natural $\langle 111 \rangle$ diamonds (macles) which are found in nature as triangular stones with perfectly smooth surfaces and therefore require no polishing or surface preparation whatsoever. Small disturbances in the diamond structure are easily detected through their effect on the channeled ion spectra, hence the high sensitivity of this technique to implantation damage in diamond.

The damage in room-temperature-implanted diamond is revealed as a clear peak in the backscattered particle spectrum at a depth which corresponds closely to the implant range. This peak grows with implantation dose until it reaches the random level. Further implantation leads to a broadening of the damage peak which, at high doses, extends from the crystal surface to a depth of roughly $R_p + \Delta R_p$. Much of this damage may be attributed to graphitization. Typical channeling spectra of implanted diamond [8.23] are shown in Fig. 8.10 in which the build-up of damage as a result of 350 keV ^{121}Sb ion implantation into diamond is evident. In this respect, diamond damage induced by heavy ion implantation closely resembles implantation-damaged Si and other semiconductors which exhibit point-defect related blockages of the channels.

The damage in ion-implanted diamond has also been observed following light ion implantation. Using proton channeling, *Derry* and coworkers [8.24, 25] have observed that implantation of 1 MeV He$^+$ into diamond leads to damage which looks different when probed along different crystalline axes in the diamond crystal. The damage is therefore not isotropic and the carbon interstitials generated by the He$^+$ implants seem to occupy a preferential site, possibly

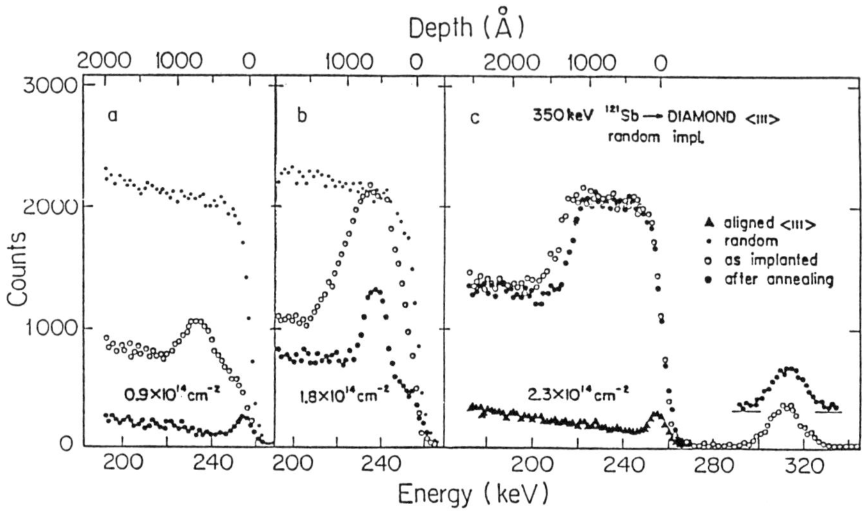

Fig. 8.10. Random and aligned energy spectra of backscattered protons from Sb-implanted (350 keV) diamonds showing the effect of the damage and of its subsequent annealing (1150°C for 1 h) for different implantation doses: (**a**) 0.9×10^{14} cm^{-2}, (**b**) 1.8×10^{14} cm^{-2}, and (**c**) 2.3×10^{14} cm^{-2} [8.23]

the tetrahedral interstitial site. More detailed channeling experiments were carried out by *Braunstein* and *Kalish* [8.26] on natural $\langle 111 \rangle$ diamonds implanted with 90 keV carbon ions to different doses. Proton channeling performed along the $\langle 111 \rangle$, $\langle 110 \rangle$ and $\langle 100 \rangle$ axes clearly showed enhanced dechanneling behind the damage peak along the $\langle 110 \rangle$ axis which indicates the existence of some non-isotropic damage in the diamond structure. Defects which may give rise to such an anisotropy could be the small diamond platelets observed in the RHEED studies of *Vavilov* [8.2] to lie parallel to the (111) planes or carbon self-interstitials which form $\langle 111 \rangle$ dumbbell configurations.

The implantation-induced damage in diamond has been investigated by *Beserman* and coworkers using the ESR technique [8.27, 28]. This technique is sensitive to both the number of unpaired electron spins via the size of the ESR signal (and hence to the total damage in the crystal) and to the nature of the dangling bond, via the position at which the resonance appears (i.e., the characteristic g-value) (Sect. 5.6). The ESR results for ion-implanted diamond identify several regimes, depending on implant dose and mass: for low doses (e.g., for antimony implantation to a dose of $\phi \sim 5 \times 10^{13}$ cm^{-2} [8.23] and for nitrogen implantations to a dose of $\phi \sim 5 \times 10^{14}$ cm^{-2}) [8.28], isolated diamond bonds are broken resulting in point defects. Beyond a certain critical high dose value, the ESR studies show that the implanted region turns into a graphitic material. This critical dose for Sb is $\phi > 5 \times 10^{15}$ cm^{-2}, and for N is $\phi > 10^{16}$ cm^{-2} [8.28].

The transformation of diamond under the influence of 1 keV Ar$^+$ irradiation has been studied in situ under UHV conditions by secondary electron emission spectroscopy (SEE). Using the SEE signatures for diamond, HOPG, amorphous carbon (a–C), glassy carbon and amorphous diamond (Sect. 5.3), *Hoffman* et al. [8.29, 30] have shown that a dose of 7×10^{14} Ar^+/cm^2 at 1 keV is sufficient to amorphize the diamond surface and to produce a surface layer which they suggest is predominantly sp^3 bonded, but without any long-range order. Higher doses result in the emergence of SEE spectra typical of a sp^2 bonded structure suggestive of glassy carbon. From these measurements it appears that low energy ion beam irradiation of diamond ultimately results in the production of a disordered sp^2-rich structure via an intermediate amorphous sp^3-rich state.

The results of the ESR studies, which essentially probe the number of broken bonds, the results of the channeling experiments which are sensitive to dislodged atoms, and the results of electrical conductivity and SEE measurements are all in agreement and thus give a consistent picture of implantation-induced damage in diamond.

8.4 Damage Annealing and Implantations at Elevated Temperatures

The transformation of sp^3 to sp^2 bonds and the creation of carbon interstitials and vacancies which accompany the slowing down process of heavy ions in diamond disrupt the metastable equilibrium of the diamond phase. Hence there is a tendency for damaged diamond to "tip over" to the thermodynamically stable form of carbon, i.e., graphite, upon addition of energy to the system. This energy may be in the form of heat or in the form of kinetic energy delivered to the lattice atoms due to the "thermal spike" associated with the damage cascade created during the slowing down process. Indeed damaged diamond was nearly always found to transform into graphite when annealed or when implanted beyond a certain critical dose.

As already mentioned in the previous section, implantation of a given ion to a given dose, when carried out at elevated temperature ($50°C < T_i < 500°C$), has two marked effects on the diamond — its electrical conductivity rises sharply for $T_i \gtrsim 50°C$ (Fig. 8.5) and its volume expands (Fig. 8.6). These observations have been explained by *Prins* et al. [8.14, 31] to be due to the different temperatures at which interstitial and vacancy diffusion set in for diamond ($\sim$ 50°C and $\sim$ 500°C, respectively), as is already discussed in Sect. 8.2.

More direct information on the nature of the defects formed at elevated temperature implantations was deduced by *Braunstein* and *Kalish* [8.26] from channeling experiments on $\langle 111 \rangle$ diamond implanted with 90 keV C ions at different temperatures ranging from room temperature to 1450°C. The dose employed was 10^{16} cm^{-2}, a dose which greatly exceeds the critical dose beyond which the implanted region seems amorphous, as far as RBS channeling is concerned, and can no longer be annealed from room temperature implantations, as will be described below. Figure 8.11 shows the RBS channeling results for implantations at 300°C, 650°C, 950°C and 1450°C. After implanta-

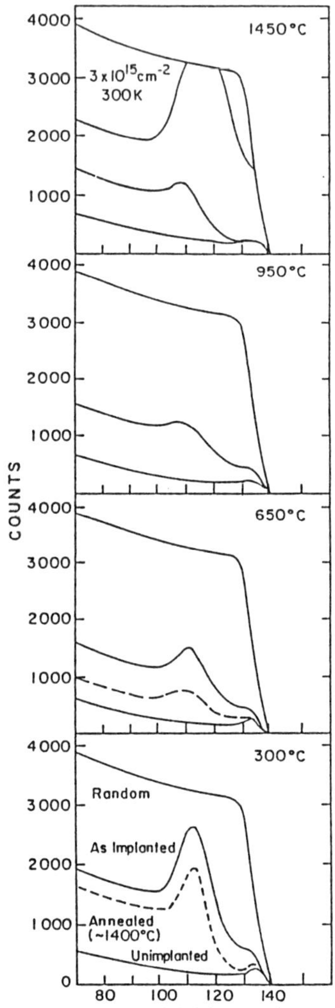

Fig. 8.11. RBS-channeling spectra of ⟨111⟩ diamond implanted with 90 keV ^{12}C ions ($1 \times 10^{16}cm^{-2}$) at various implantation temperatures (T_i), showing the dependence of the residual damage on T_i noted in the upper right hand corner of each figure. Random and channeling spectra taken for unimplanted (virgin) diamonds as well as channeling spectra obtained after annealing at $\sim$ 1400°C (dashed curves) are also shown [8.26]

tion at 300°C, a typical damage peak is observed indicating that the damage is mostly of the obstruction-type (point defects), and post-implantation annealing at 1400°C (1 h, 10^{-5} Torr) does not remove this damage peak, probably due to the graphitization of a thin buried damaged layer. After implantation at 650°C, the damage peak is reduced, but the dechanneling is enhanced. The heating of diamond to 650°C during the implantation is thus sufficient to maintain the damage level below the graphitization threshold, but cannot inhibit the formation of distortion-type (extended) defects. After further annealing at 1400°C (dashed lines in Fig. 8.11) the damage peak has almost disappeared (i.e., point defects have annealed), but enhanced dechanneling is still observed in the backscattering spectrum, indicating that those extended defects formed during the 650°C hot implantation are fairly stable. Following implantation at 950°C, the damage peak is hardly noticeable and only near the surface do some point defects appear to remain. Implantation at 1450°C causes even the near-surface region to anneal well. Dechanneling of the probing beam is, however, observed in the implanted zone, indicating the presence of the distortion-type defects in the region where the implants came to rest. For comparison, Fig. 8.11 also shows the RBS-channeling spectrum of a diamond implanted at room temperature with 90 keV, 3×10^{15} C ions/cm^2. Even though this implantation dose is only a third of that of the high temperature implantations, a wide damage peak which reaches the random level is observed.

Post implantation annealing of damaged diamond perfectly restores the crystallinity only if the damage density does not exceed a certain value. The data in Fig. 8.10 (full large circles) display the channeled spectra of 350 keV protons following annealing (1 h at 1150°C in vacuum) of diamonds implanted at room temperature with 300 keV ^{121}Sb ions to different doses. While complete annealing is obtained following implantation of 0.9×10^{14} cm^{-2}, the damage which results from an implantation to 1.8×10^{14} cm^{-2} does not disappear, but rather remains as a sharp peak at approximately the implant range. This peak may possibly be associated with a buried narrow graphitic layer. The fact that the RBS peak does not reach the random level may be due to the narrowness of the layer relative to the detector resolution.

For even higher doses (2.3×10^{14} cm^{-2}), no damage removal occurs whatsoever, other than a slight epitaxial growth from the diamond/damage interface. The 'annealed layer' for this, and higher dose implantations exhibits many graphitic properties including etchability by acids which are known to attack graphite and not diamond, a fact that has been used by *Braunstein* and coworkers to remove thin diamond layers in a well controlled way [8.32].

The annealing of diamond implanted under conditions which damage it below the critical level beyond which annealing is no longer possible (i.e., 90 keV C, 1×10^{15} cm^{-2}), has been studied by channeling experiments [8.26]. The results of channeling measurements indicate that complete damage removal can be obtained only when extremely high annealing temperatures (1450°C) are employed [8.26].

In contrast to the results of [8.26], which apply to the annealing of ⟨111⟩ diamond implanted at room temperature, *Liu* et al. [8.33] have studied the annealing behavior of ⟨100⟩ diamond implanted at 77K and subsequently annealed. In this work 200 keV carbon ions have been implanted into ⟨100⟩ natural diamonds to doses ranging from 10^{15} to 3×10^{15} cm^{-2}. The implanted diamonds have been subjected either to rapid thermal annealing (1100°C for two minutes) or to isochronal annealing for 1 h at temperatures ranging from 450 to 900°C. RBS channeling analysis has clearly shown that: (a) prolonged annealing in a

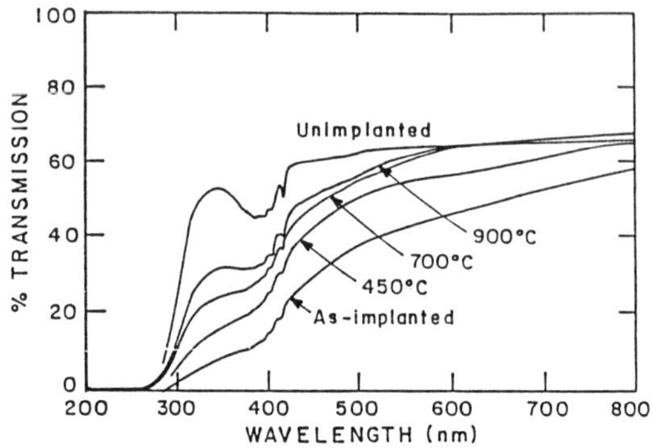

Fig. 8.12. Optical spectra for a type Ia diamond implanted with 200 keV C ions to a dose of 1.25×10^{15} cm^{-2} following isochronal annealing for 1 h at the temperatures indicated [8.33]. The spectrum for the unimplanted sample is shown for comparison

furnace yields superior results to those obtained following rapid annealing, (b) an anneal at 900°C for 1 h is sufficient to remove the damage following cold implantations, and (c) there is a critical dose for amorphization of diamond which is around 2×10^{15} cm^{-2} for the cold C implantations employed in [8.33]. Optical transmission measurements covering the wavelength range of 200-800 nm have verified the annealing behavior deduced from the RBS channeling experiments, as shown in Fig. 8.12, but indicate that some damage, undetectable by channeling, remains even after annealing at 900°C for 1 h.

Very similar conclusions were drawn by *Lee* and coworkers [8.34] from their studies of hot implantations into diamond by means of electron spin resonance (ESR) techniques. They found that the ESR signal attributed to amorphous carbon (which probably corresponds to what is identified in channeling experiments as point defects), is reduced as the implantation temperature is increased; however implantation above 600°C results in the formation of two new ESR-active centers labeled A.5 and A.6. It was proposed in that work that the defect responsible for the A.5 feature in the spectrum could be a distorted hexavacancy ring, and that some other vacancy clusters were responsible for A.6. Such multivacancy clusters can very well account for the typical dechanneling behavior observed in hot-implanted diamonds. ESR was also employed by *Braunstein* et al. [8.23] and by *Teicher* and *Beserman* [8.28] in their study of the annealing of Sb and N implanted diamond. In that work type Ia diamonds were implanted [8.28] with Sb ions at 300 keV at different doses ($4 \times 10^{12} - 2 \times 10^{15}$ cm^{-2}) and their ESR spectra were taken before and after annealing at 1000°C for 1 h in vacuum. Large changes in both line intensity and line width were observed as a function of implantation dose; these changes are much diminished upon annealing as seen in Figs. 8.13 and 8.14. Of particular interest is the steep variation in both line shape and intensity which occurs at an Sb dose of $\sim 2 \times 10^{13}$ cm^{-2}. This dose is one order of magnitude below that at which the sharp rise in conductivity, attributed to the formation of graphitic bonds [8.5], sets in (which

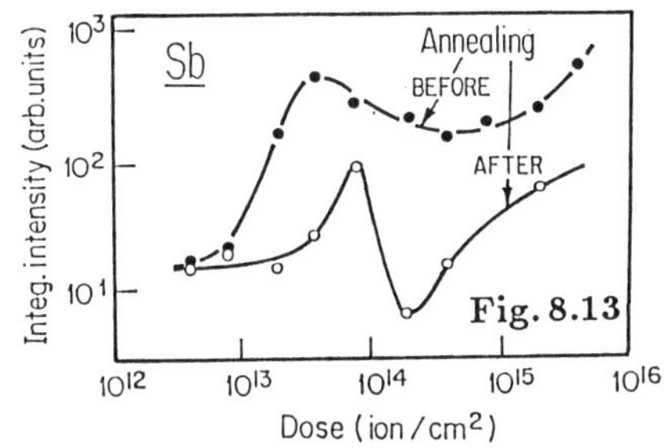

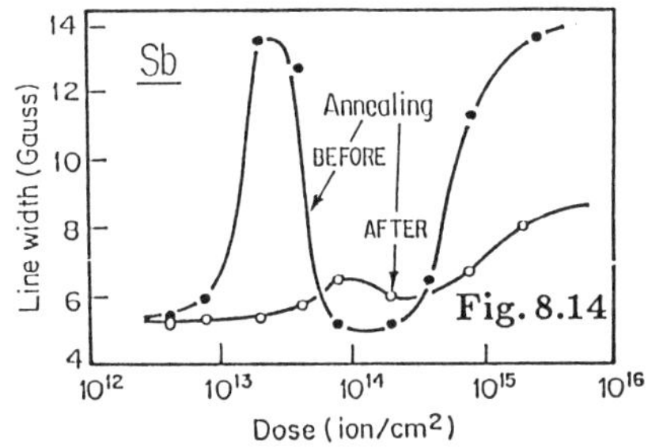

Fig. 8.13. Integrated intensity of the ESR signal in ion implanted diamond as a function of the antimony implantation dose before and after annealing [8.28]. The Sb implantation was at 300 keV and the annealing was one in vacuum at 1000°C for 1 h

Fig. 8.14. Width of the ESR signal in ion implanted diamond as a function of the antimony implantation dose before and after annealing [8.28]. The Sb implantation was at 300 keV and the annealing was done in vacuum at 1000°C for 1 h

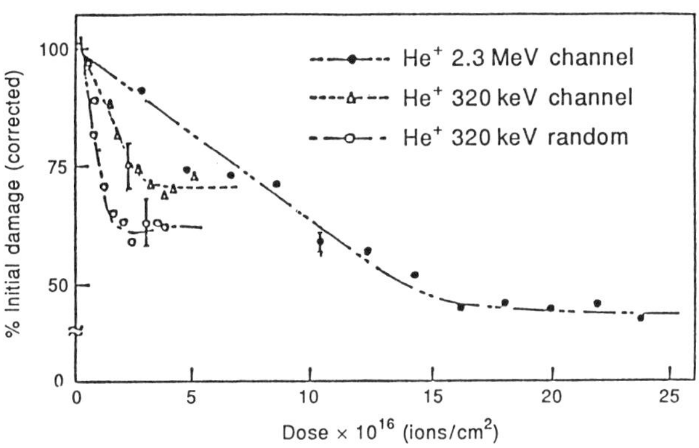

Fig. 8.15. Percentage of initial damage (by 300 keV Sb ions, $\phi = 10^{14}$ cm^{-2}) remaining in the first 1000Å of $\langle 111 \rangle$ diamonds as a function of the net annealing beam dose for He and H ions incident along channeling and random directions [8.35]

is at $\sim 2 \times 10^{14}$Sb/cm^2 for similar implantation conditions, a dose at which another irregularity in the ESR line width is noted). Interestingly, in contrast to the channeling (Fig. 8.10) and to chemical etching data which show that at this dose annealing leads to graphitization of the damaged layer, no dramatic alterations in the ESR spectra are noted to occur as a result of annealing.

Despite the thermodynamic instability of the diamond phase, it has been shown by *Adel* and coworkers [8.35] that light ion-beam induced annealing occurs in diamond when damaged below the same critical dose beyond which its structure cannot be restored even by thermal annealing. $\langle 111 \rangle$ diamonds, damaged by implanting 300 keV Sb ions to a dose of 10^{14} cm^{-2}, were subjected to high doses of 2.3 MeV He$^+$ or 320 keV H$^+$ beams impinging on the diamond under either channeling or random incidence conditions (Fig. 8.15).

Damage shrinkage has clearly been observed by these channeling experiments, and this shrinkage has been shown to be caused predominantly by nuclear, and not electronic collisions. In that respect, diamond does not differ from other, covalent crystals for which ion-beam induced annealing has been observed. (See for example *Elliman* et al. [8.36].)

8.5 Electrical Doping

While studies of diamond implanted with carbon or noble gas ions are aimed mainly at the basic understanding of the damage mechanism and the structural transformations experienced by the disrupted crystal, implantation and annealing studies of diamond implanted with ions which are potential donors and acceptors in this unique wide band gap semiconductor have very practical implications. The great advantages of a semiconductor device based on diamond relative to existing devices based on smaller gap materials have long been realized [8.37]. These advantages have recently been summarized by *Geis* and coworkers [8.38, 39] who illustrate how the extremely high electron velocities, the high breakdown fields and the high thermal conductivity of diamond can make unique high speed devices with high power capability and resistance to radiation damage.

In order to achieve p- or n-type conductivity, group III or V impurities must be introduced into the diamond and these impurities must be located on specific sites in the lattice (usually substitutional) where they may be electrically active. Since diffusion of impurities in diamond is extremely slow, and since our knowledge of synthetic diamond growth incorporating the desired impurities is still in its infancy, ion implantation has been attempted by many researchers as a means of forcing the required dopants into the crystal. It should be mentioned that an alternate approach to the introduction of dopants into diamond by ion implantation is through homo-epitaxial growth of diamond layers with dopant atoms incorporated during growth. Success with this approach has recently been reported; however this subject is beyond the scope of the present review [8.39]. The advantage of ion implantation over other doping techniques as a method of choice for the introduction of donor impurities (Li, N, Na, and P) into diamond have most recently been predicted by the *Bernholc* group [8.40]. The implantation process in diamond, as has already been shown, is accompanied by much damage, which may by itself give rise to electrical conductivity. Thus, appropriate annealing conditions must be found to remove the lattice damage and to drive the implants to the desired lattice sites. Since thermal annealing of implantation-damaged diamond may also lead to graphitization of the damaged region and hence to a sharp rise in its electrical conductivity, the procedure for obtaining conductivities due only to the dopants and not to lattice damage is a very complex one, and many misleading results have been published on this subject.

Attempts to implant diamond and subsequently anneal out the implantation-induced damage have centered around a limited number of potential donor or acceptor ions. As already pointed out (Sect. 2.1.6), p-type diamonds (type IIb) are found in nature with boron as the acceptor ion. Nitrogen is also found in natural diamond, but usually not in an electrically active form. Most studies have thus aimed at either "repeating what nature can do", i.e., achieve p-type doping through B implantation, or by trying to obtain n-type conductivities through implanting potential donor ions: Li, P, As or Sb [8.41]. Channeling studies were performed in order to optimize the annealing conditions and to study the location of the implants in the lattice [8.42]; hyperfine interaction experiments (to be discussed in Sect. 8.6) were made in some selected cases to investigate the immediate surroundings of the implants; electrical measurements were carried out to search for the desired electron or hole conductivities [8.26], and, in a very few cases, devices were realized, as will be described below.

Pioneering work attempting to utilize the semiconducting properties of diamond were already carried out in the seventies by scientists from the Lebedev Institute in Moscow and were summarized in a series of review articles by *Vavilov* [8.37, 43] and by *Vul* [8.44]. Many of the topics touched upon by these early researchers have since been repeated under improved measurement conditions, thereby giving rise to more accurate experiments and more detailed interpretations. Three different approaches have been taken by various groups to overcome the interfering effects caused by possible annealing-induced graphitization: hot implantation [8.42], annealing followed by graphite removal [8.26], and C^+ co-implantations [8.31, 45].

The structural recovery of diamond implanted at high temperatures ($>$ 1000°C) has been studied by RBS channeling experiments [8.42]. These studies have clearly shown that the diamond lattice remains nearly undamaged, even following high dose implantations (10^{16} cm^{-2}), as long as the diamond is kept at high enough temperatures during implantation ($T_i > 1000$°C). Even though some defects do remain in the implanted region for such high temperature implantations, these are extended defects, and are completely different from the point defects observed after room temperature implantation or after room temperature implantation followed by high temperature annealing, as inferred from RBS channeling experiments [8.23].

The channeling technique has also been used by *Braunstein* and *Kalish* [8.42] to study the lattice locations that the implants occupy in the diamond lattice, using the appropriate probing method to locate the impurities [e.g., RBS for Ge and Sb, PIXE for P, and nuclear reaction analysis (NRA) for ^{6}Li and for ^{11}B (Sect. 4.4)]. The results of this study [8.42] clearly show that Sb cannot be driven onto substitutional sites even under high temperature implantations, nor does the P^+ implant seem to occupy predominantly substitutional sites, as they should in order to be active donors. Therefore both P and Sb are unlikely to act as effective donors in diamond, regardless of the annealing method employed. This observation may be explained by the different sizes of the P and Sb atoms relative to carbon. Similar results on the lattice site location deduced from channeling experiments and the lack of electrical activation for P^+-implanted diamond were also reported by *Davidson* et al. [8.46]. In contrast to the P implants, the location of Li implants in diamond (Fig. 8.16) is

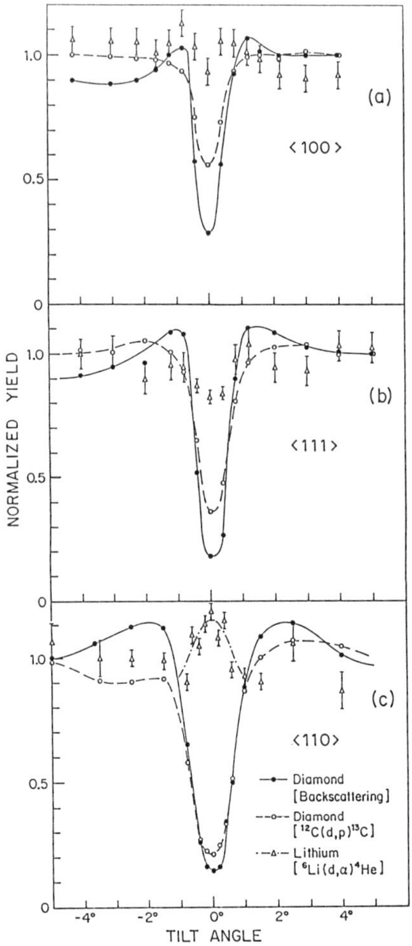

Fig. 8.16. Angular dependence of the yield of the α particles from the ^{6}Li(d,α)^{4}He reaction (*triangles*), of protons from the ^{12}C(d,p)^{13}C reaction (*open circles*), and of backscattered deuterons (*full circles*), all measured simultaneously on a ^{6}Li implanted diamond. Parts (a), (b) and (c) correspond to angular scans around the ⟨100⟩, ⟨111⟩ and ⟨110⟩ axes, respectively. Flux peaking is revealed in the angular scan of the α particles along the ⟨110⟩ axis indicating the blockage of this channel by the Li implants [8.42]

most encouraging, as substantial numbers of Li ions were found to occupy interstitial sites, following implantation (90, 60, and 40 keV to a total dose of $2.3 \times 10^{16}/\text{cm}^2$) into diamond held at 1000°C. Li atoms residing on these sites are predicted, by analogy to the doping of Si, to be electrically active donors. Indeed Hall effect measurements by *Vavilov* and coworkers [8.47, 48] have shown n-type conductivities in Li-implanted diamond, with mobilities on the order of 1000 cm^2/(V s) and an activation energy of 0.10 eV.

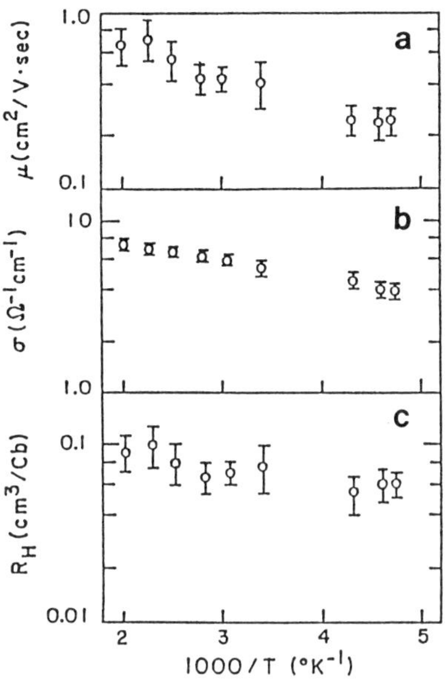

Fig. 8.17. Temperature dependence of the (**a**) Hall mobility, (**b**) electrical conductivity, and (**c**) Hall coefficient of a type IIa diamond doped with boron ions by the implantation-etching procedure [8.26]

Braunstein and *Kalish* [8.26] have developed a method to overcome the disturbing effects of graphitization that accompany the high temperature annealing of heavily damaged diamond. Their experiments were carried out on diamond implanted with 40 keV B ions to a high dose (10^{16} cm^{-2}), exceeding the critical dose above which graphitization occurs as a result of annealing. Following high temperature (1400°C) annealing of the sample in vacuum, the damaged region, which for such implantations extends from $\sim (R_p + \Delta R_p)$ all the way to the surface, was transformed into graphite. This graphite layer could be chemically removed [8.32]. Nuclear reaction analysis (NRA), combined with channeling has shown that a thin surface diamond layer, which remained untouched by the chemical etchants, still contained appreciable amounts (10^{20} cm^{-3}) of B dopants, residing in an annealed diamond lattice on substitutional (electrically active) sites [8.26]. This doped layer indeed exhibited high electrical conductivity, on the order of 10 Ω^{-1} cm^{-1}, in contrast to an identically treated diamond which was implanted with C ions. This latter control sample exhibited, after graphite removal, the extremely low conductivity ($< 10^{-9} (\Omega \text{ cm})^{-1}$) of the original undoped diamond, in sharp contrast to the similarly treated B-implanted sample. Measurements of the Hall effect, and of the temperature dependence of the conductivity for the B-implanted diamond, shown in Fig. 8.17 prove that the conductivity is caused by holes and that it can be characterized by a low (0.02-0.06 eV) activation energy. Such low activation energies are common to heavily B-doped synthetic diamonds. It should be mentioned that lower dose implantations of B and C ions (3×10^{14} cm^{-2}, 40 keV) have both yielded, after 1400°C annealing, the extremely high resistivity of the virgin stones, indicating that no electrical activation nor any graphitization had occurred under these conditions [8.26].

The extensive work of *Prins* on the temperature dependence of the volume expansion and of the electrical conductivity of C-implanted diamond [8.11] has clarified the important role that the implantation temperature plays in the nature and in the spatial distribution of the defects in the implantation affected layer. For implantations at low temperatures (defined as T_i < room temperature) both interstitials and vacancies are "frozen in" during the implantation [8.11]. If, according to *Prins* [8.31], dopant atoms, like B, are co-implanted with C ions into the same volume at this low temperature, and if the system is then rapidly annealed, there should be a chance for the implanted foreign atoms to compete with the self-interstitials in the filling of vacancies. The result may thus be that an appreciable number of dopant atoms will reside on substitutional sites in an undisturbed diamond lattice, hence becoming electrically active. This idea was tried out by *Prins* [8.31] who implanted C and B ions into type IIa diamond held at LN_2 temperature using different carbon to boron dose ratios so that the total dose was always 5×10^{16} cm^{-2}. The implant energies were chosen to create a fairly uniform damage profile extending from the surface of the diamond to a depth of 2000Å. The diamond was then rapidly annealed by dropping it from its holder which was at LN_2 temperature onto a hot plate held at 500°C. Subsequently, the diamond was further annealed by placing it in a furnace held at 1200°C for 2 h. The properties of diamonds treated in this way were characterized both optically and electrically. Most convincing are the results of the sheet resistivity versus inverse temperature of the B-implanted layers for different B fractions in the B+C implantation (Fig. 8.18). While the specimen implanted with pure B showed a very low sheet resistance with a slow temperature dependence corresponding to an activation energy of 0.02 eV (similar to that measured by *Braunstein* and *Kalish* [8.26] for a heavily-doped B layer), increases in resistivity and increased activation energies were observed with decreasing B fraction. *Prins* identified several slopes in the ρ versus $1/T$ curves which he correlated with different conduction mechanisms. In particular one of the slopes of the 70% line which corresponds to an activation energy of 0.37 eV, known to be that for substitutional B in type IIb diamonds, was taken by *Prins* as proof that substitutional B acceptors are present in this particular sample [8.31].

Sandhu et al. [8.45] picked up on *Prins*' ideas and have performed dual implantation experiments very similar to those of *Prins* [8.31]. They, too, have co-implanted diamond held at 77 K with C^+ and B^+ ions, so that the C vacancy and B implant profiles overlap. Following the implantations, isochronal annealing (300° – 900°C in vacuum) was performed, and the samples were evaluated by taking optical absorption curves in the visible and in the infrared spectral regions. The gradual annealing of the radiation damage is indicated by the decreasing absorption of the GR-1 band, associated with C vacancies, while the incorporation of B onto substitutional sites in the diamond lattice is manifested

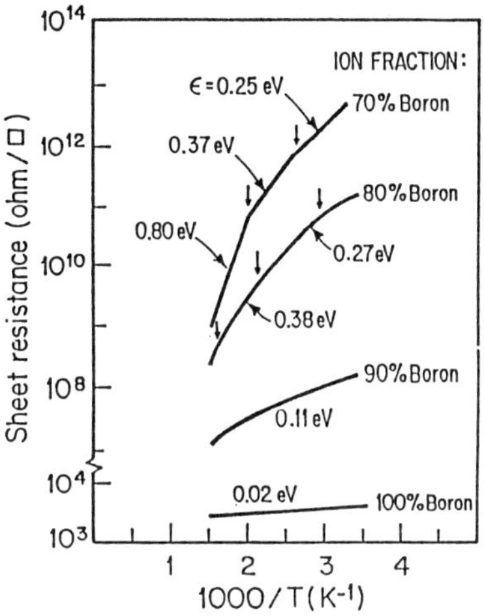

Fig. 8.18. Sheet resistance as a function of inverse absolute temperature for some of the carbon and boron co-implanted layers obtained in diamond. Activation energies are labeled on the curves for the various boron ion fractions [8.31]

by the increased absorption of the lines at 2957 cm^{-1} and at 2925 cm^{-1}, which are identified with substitutional B.

More recently the same group [8.45, 49] complemented their optical characterization with electrical measurements on similarly implanted and annealed type IIa diamonds. Their results verify those of *Prins* [8.31] in that they also show that B can be electrically activated to a degree which depends on the relative fluence of boron and carbon ions, with activation energies closely resembling those of natural *p*-type diamonds.

8.6 Impurity-State Identification

A good deal of effort has recently gone into more detailed studies of the site symmetries of ion-implanted materials in general, and diamond in particular, using a variety of experimental techniques. These are summarized in this section.

A very sensitive way to determine the exact site that an impurity occupies in a crystal and the perfection of its immediate environment is to measure the hyperfine-interactions that the impurity nucleus experiences. The two major techniques for this are to utilize the Mössbauer effect (ME) or the perturbed angular correlation (PAC) method. Both are, however, limited to a few selected nuclear probes which have the required special nuclear properties, as previously described in Sect. 5.7.

The first PAC measurement on impurities implanted into diamond was performed by *Kalish* and coworkers [8.50], who implanted low doses ($< 3 \times 10^{13}$ cm^{-2}) of ^{111}In ions at 350 keV into diamond as well as into graphite and into graphitized diamond. The samples were then annealed up to very high temperatures (2100 K) in high vacuum, and both PAC and radioactivity (In) loss measurements were performed. Whereas the implanted graphite samples lost all radioactivity upon annealing at temperatures exceeding 1450 K due to In out-diffusion, the diamond retained all In up to the highest temperature that could be reached (2100 K). Nevertheless the PAC spectra looked rather similar for all implantation and annealing procedures employed. These PAC spectra are characterized by the absence of distinct features (i.e., no well-defined unique defects present in the vicinity of the probe nuclei), and all spectra exhibit a very fast decay of the anisotropy, indicating that the In atoms reside in environments where they experience a variety of very strong field gradients. No trace of In nuclei residing on "unperturbed" perfect substitutional sites could be detected, either in diamond or in graphite [8.50]. The conclusion drawn from this work was that because of the very dense damage cascade caused by a heavy ion implant, and because of the large size of the foreign atom (as compared to the typical "nearest neighbor" distances in diamond or graphite), there exists an intrinsic limitation on doping diamond by heavy ion implantation regardless of annealing. A similar experiment was carried out by *Appel* et al. [8.51], with the difference being a shallower (65 keV) ^{111}In implantation than that of [8.50], and the implantation was done into heated (900 K) diamond. Even though *Appel* et al. detected a small fraction (5%) of ^{111}In nuclei occupying a specific site in the diamond crystal, they too could not find appreciable amounts of implants residing on unperturbed lattice sites where they might be electrical donors. Similar results were obtained by the same group [8.52] when implanting ^{181}Hf into diamond.

More recently *Connell* and coworkers [8.53] used recoil implantation of ^{19}F into a pure type IIa diamond, into other type diamonds, and into graphite. Their diamonds were held at different temperatures: LN_2 (where vacancies and interstitials are immobile), RT and 200°C (where only interstitials are mobile) and 550°C (where both vacancies and interstitials may be mobile). None of the experiments yielded purely unperturbed sites for the F^+ implants; however, a slightly distorted tetrahedral interstitial site predicted to be favored by F^+ ion incorporation into diamond [8.54] could perhaps have been identified.

Some experiments have been published in which Mössbauer probe nuclei (mainly ^{57}Co and ^{133}Xe) were implanted into diamond and graphite. Of particular interest has been the work of the *Leuven* group [8.55, 56], and the more detailed studies of the *De Potter* and *Langouche* group [8.57, 58] and the *Groningen/Cracow* groups [8.59, 60] who implanted ^{57}Co into diamond and graphite under a variety of conditions. Their results, as well as those published by *Langouche* [8.61] using the Mössbauer ^{133}Xe nucleus implanted into diamond, all seem to indicate that upon annealing some implants can be driven into cubic sites in diamond. Furthermore, following implantation at low (LN_2) temperatures, the measured Debye-Waller factor differs greatly from that found for room temperature implantation; the large probe atoms experience high internal pressures; and in most cases the immediate environments of the implants in diamond closely resemble those of graphite.

8.7 Electronic-Device Realization

The study of implantation damage in diamond, its annealing and the regrowth of the diamond structure are not only of purely scientific interest. The possibility of utilizing the semiconducting properties of diamond in order to realize electronic devices is of great technological importance. This requires, however, the possibility of doping diamond *p* or *n* type, with ion-implantation (in analogy to Si) being a method of choice to achieve this. There are thus important practical implications for the study of ion-implantation in diamond and of the damage-related anomalies, with the goal of realizing reliable working electronic devices in this unique wide band-gap semiconductor. In this respect, unfortunately, not much progress has yet been made, though several reports have been published on devices which showed promise. However, these results presumably could not be reproduced, since no follow-up of those reports have appeared in the literature, nor have products utilizing these devices appeared on the market. (Nevertheless, quite a few patents on these concepts have been filed!).

As early as 1973 some rectifying effects were found by *Glover* [8.62] in Schottky devices based on natural and synthetic boron-containing diamonds. The extensive work carried out in the U.S.S.R. in the 1970s on the topic of implanting diamond to obtain devices is reviewed in the papers of *Vul* [8.44] and *Vavilov* [8.37]. Both reviews show *I-V* curves for *p-n* junctions obtained in diamond by either B and P, or B and Sb implantations. The best results were obtained following annealing at 1400°C. Some of the devices even showed a photo-response to irradiation with UV light. However, no follow-up of these early "successful" results has appeared in literature. Furthermore, in the light of extensive studies on the basic science relevant to the implantation and annealing of implanted diamond, reviewed in the previous sections, it is possible that these early researchers were misled by other implantation- and annealing-induced phenomena (e.g., graphitization, etc.).

More recently *Prins* [8.63] reported on the successful realization of a transistor in diamond. *Prins*, being aware of the difficulties in electrically activating dopants implanted in diamond by thermal annealing, decided to utilize the *p*-type conductivity that nature has provided in natural type IIb diamonds, together with the *n*-type conductivity exhibited by damaged diamond [8.43]. The transistor thus fabricated (Fig. 8.19) consists of a type IIb diamond, the surface of which was heavily damaged by a sequence of C^+ implantations, designed to create a uniform damage layer, extending from the surface to a depth of 0.3 μm. Such a heavily implanted layer was shown by *Hauser* [8.8] to conduct by electron hopping. A thin W wire (3.2 μm thick) was used to mask a line from this implantation, leaving the underlying material *p*-type. This arrangement to which simple contacts were made by pressing Au balls onto the *n*-type layers and silver to the *p*-type base, demonstrated typical diode and transistor characteristics as shown in Fig. 8.20. The current amplification factor of this device was only ~ 0.11, possibly due to its unfavorable geometry. Unfortunately, this unique device has never been reproduced or improved, either by *Prins* [8.63] or by anyone else.

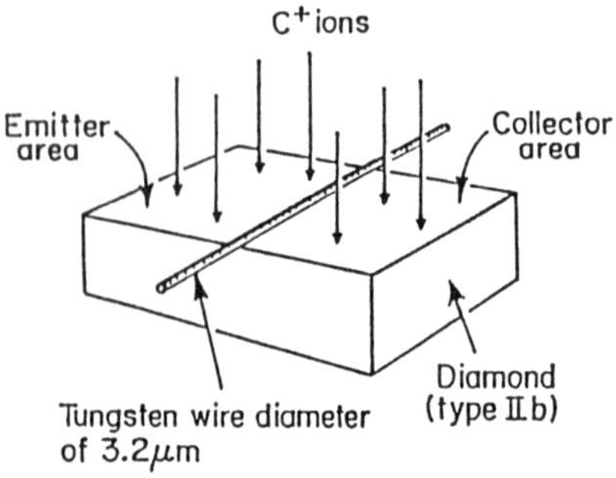

Fig. 8.19

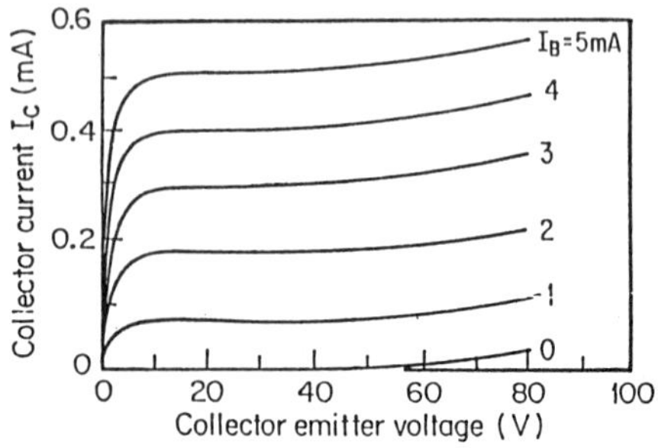

Fig. 8.20

Fig. 8.19. Schematic illustration of the masking of the base area during carbon ion implantation, used in the realization of a field effect transistor in a type IIb diamond [8.63]

Fig. 8.20. *I-V* characteristics of a diamond field effect transistor produced as shown schematically in Fig. 8.19 [8.63]

Though not exactly related to the subject of the present review, which deals with ion implantation, it is worthwhile to end this section on a positive note regarding the rapidly growing field of diamond film deposition by CVD and other techniques, and the promise of these techniques in electronic device fabrication in diamond. These techniques have allowed researchers to grow epitaxial diamond films onto single-crystal diamond with dopant atoms incorporated in the films during growth. This procedure obviously circumvents the damage annealing problem which is inherent in the implantation process and enables the construction of *p-n* structures in the same way as is sometimes used in conventional Si technology. Most recently, *Geis* [8.39] has described a device produced by growing single crystal doped diamond layers on a diamond substrate. This experimental device indeed seems to exhibit many of the promising features expected in semiconducting devices based on diamond.

8.8 New Materials Synthesis

Very high dose implantation ($> 10^{17}$ cm^{-2}) can introduce large concentrations of foreign (implanted) atoms in the substrate, centered around R_p, so that new materials can sometimes be synthesized. This procedure, which nowadays is commonly used to build buried SiO_2 layers in Si by very high dose O implantation into Si, has also been applied to diamond in an attempt to stimulate SiC or diamond growth by Si or C implantations, respectively, into diamond.

Most efforts to obtain a SiC layer in diamond by Si implantation have been done by the *Lebedev* group [8.64, 65] who have employed X-ray diffraction, IR absorption and ion scattering techniques to evaluate their materials. In their work, 30–40 keV ^{28}Si ions were implanted into diamond to very high doses (up to 4.4×10^{17} cm^{-2}) and the composition and structure of the Si+C mixed layer so formed was studied following different annealing procedures. The stoichiometric ratio of 50 atomic percent which is required to synthesize the SiC phase could indeed be reached by such implantations. RBS measurements showed, for example, that the implanted Si layer produced by 4.4×10^{17} Si/cm^2 at 40 keV contained as much as 59% Si in diamond near R_p [8.64, 65]. Detailed studies of the structure of Si-implanted diamond and the phases formed in it have been carried out by *Krasnopevtsev* et al. [8.66] who have performed differential RHEED experiments. Diamonds, subjected to 30 keV Si implantations to different doses (4.8×10^{14} cm^{-2} – 2×10^{17} cm^{-2}) and to different annealing temperatures (up to 1200°C) have been analyzed by the RHEED technique following successive layer removal by etching in $HClO_4$. The RHEED pictures have shown the existence of small, randomly-oriented 30–50Å crystallites, identified as having the β-SiC phase. These were found to be mostly concentrated in the top 300Å of the implanted diamond, a thickness which coincides with the projected range of the Si implants. It is claimed in [8.67] that graphitization, which based on today's knowledge should take place at the high dose implantation levels employed in [8.67], has been prevented by the existence of the tetrahedral Si-C bonds. Nevertheless, the fact that the implanted layer could be attacked by acids, which is not the case for both diamond and SiC, raises the question of whether the phases reported [8.67] were indeed SiC embedded in diamond. The opposite approach (i.e., the formation of SiC in Si by C ion implantation into Si) has also been attempted, and this approach has yielded, following annealing, a layer consisting of nearly 100% of the β-SiC phase. The implantation of Si into diamond has turned out to be much less effective, yielding at most 15% of the desired SiC phase [8.64].

A patent by *Nelson* claiming that diamond growth can be achieved by high dose carbon ion implantation into heated (400°–1200°C) diamond [8.68] has generated enthusiasm, and attempts to increase the size of natural diamonds by this method have followed. The diamond layer grown by this method was, however, found to be defective, containing large amounts of extended defects which manifested themselves through enhanced dechanneling of the RBS probe ions [8.69]. These defects were attributed [8.67] to the agglomeration of migrating point defects into a dislocation array during the hot implantation. Auger electron spectroscopy verified that the material grown by high dose C^+ implantation consisted predominantly of diamond. While the mechanical and chemical properties of the as-grown diamond are identical to those of single-crystal diamond, their optical properties indicate the presence of defects which give the implanted diamond a yellowish or light brownish appearance. Therefore this method of increasing the diamond size has up till now had no real value to the gem stone industry.

However, growth in the understanding of implantation-induced damage in diamond and the annealing of this damage has in the last few years been quite remarkable. It is therefore conceivable that reexamination of some of the early attempts to synthesize diamond-based materials might yield results superior to those obtained nearly 10 years ago, when the understanding of the relevant processes was not as complete as it is today.

8.9 Improving Mechanical Properties

It is well known that the mechanical properties of steels and other superhard materials can be improved by ion implantation, the major implant species being nitrogen [8.70]. It is also known that type I diamonds, which contain as much as 0.25% nitrogen, exhibit superior hardness. Therefore, it is not surprising that improvement of the mechanical properties of diamond by ion (mostly nitrogen) implantation has been attempted. The goal of these studies has been to enhance the frictional and scratch resistance of diamond, as well as to improve its wear behavior and its machining performance as a cutting edge. *Hartley* [8.71] implanted (at room temperature) various diamond plates, tips, and tools with N^+, B^+ and C^+ ions at 100–300 keV to doses of $\sim 3-5\times10^{15}$ cm^{-2}, and exposed these specimens to a series of scratch and wear tests under a variety of load conditions. Interestingly, some marked improvements were found in the mechanical properties of N^+-implanted diamond, as compared to non-implanted or identically C^+-implanted diamond tools.

The most striking of those improvements in mechanical properties is the finding that a phonograph stylus in contact with a rotating TiC disc outlasted those without implantation by at least an order of magnitude [8.71]. The same was found for N^+-implanted diamond-finished cutting tools, which showed substantial improvement in their lifetime in machining acrylic plastic. A few experiments where B was implanted seem to show the same trend, in contrast to carbon implantation, which did not leave any marked effect on the mechanical properties in the implanted diamond layers. Since the wear of the diamond is believed to be due to loss of material caused by cleavage or peeling of platelets from the diamond surface, the hardening effect caused by N (or B) implantation must suppress this cleavage through both structural and chemical modifications. *Hartley* proposed that a stiffer outer layer is formed on the implanted diamond, possibly because of radiation-enhanced migration of impurities and vacancies to form stable complexes. These results were reproduced by a Chinese group [8.72] who claim that N^+ and B^+ implantations improve the wear resistance of synthetic polycrystalline diamond by a factor of ~2.

Even though the largest commercial market of diamond is in the cutting, machining and drilling tool industry, no extensive scientific work seems to have been published on either systematic investigations of the effect of ion implantation on the improvement of the mechanical properties of diamond nor on the basic understanding of these hardening effects on an atomistic scale.

It is worth noting that mechanical properties of graphitic materials, in particular glassy carbon, were also shown to dramatically improve upon ion implantation (Sect. 7.1) with N probably being the ion of choice. The question why this particular element (N) yields the best improvements in the mechanical properties of C-based materials is still an open question, the solution to which may lie in a unique chemical property of the C–N bond.

9. Implantation-Induced Modifications to Diamond-Related Materials

Two families of man-made materials which are closely related to diamond have been developed recently, and the ion-induced effects relevant to these are mentioned here. The first, which was already grown ~20 years ago [9.1], are amorphous-hydrogenated carbon (a–C:H) (or, as sometimes called, diamond-like carbon, DLC) films and the second are the recently-developed man-made thin diamond films. Both of these are grown in the form of thin coatings from a plasma which contains a mixture of carbon and hydrogen ions. In the following, ion implantation studies in DLC are briefly reviewed. As for implantation effects in diamond films, hardly any data exist on these, probably because of the novelty of the material. However, from the available data [9.2] it seems that diamond films respond to ion implantation exactly like diamond single crystals do, at least as reflected in ion-induced changes in the electrical conductivity and in the Raman spectra.

9.1 Diamond-Like Carbon (a–C:H) Films

Diamond-like carbon films are, in a sense, the analogy to amorphous hydrogenated silicon, both being amorphous materials based on a group IV element and both containing substantial amounts of hydrogen. However, the basic difference between C and Si, namely the possibility of carbon to form both sp^2 and sp^3 bonds, in contrast to Si which binds only in its tetrahedral configuration, is reflected in many of the properties of a–C:H films. Depending on the preparation method, the ratio of sp^3 to sp^2 bonds in the material can vary and, with it, also the properties of the film. Nevertheless, typical "good" DLC films are more diamond-like than graphite-like, as can be seen in Table 2.2, which compares some of the physical properties of DLC to those of diamond and graphite.

In order to grow DLC and not graphitic films, one has to circumvent the natural tendency of carbon to form stable sp^2 graphite-like bonds as opposed to sp^3 metastable diamond-like bonds. A method is therefore required to 'bypass' this thermodynamic obstacle. This is achieved by using carbon ions with sufficient kinetic energy and by film growth in the presence of H^+ ions. Virtually all the preparation methods may be categorized as either plasma discharge or direct ion beam techniques, both of which are non-equilibrium processes. This implies that the ion-surface or plasma-surface interactions may not be satisfactorily described using only macroscopic thermodynamic parameters such as gas pres-

Table 9.1. Influence of ion impact energy on the material produced in the plasma discharge of hydrocarbon gases [9.3]

Ion impact energy [eV]	Material produced
1000	dense carbon
100	dense hydro-carbons
10	polymer-like films
1	plasma polymers

sure or substrate temperature. However, on a microscopic level, the preferential growth of sp^3 bonds may be understood in terms of the dynamical evolution of the collision cascade. An energetic ion impinging on a surface causes a transient thermal spike on a picosecond time scale (10^{-12} s) followed by an ultra-fast quenching, thus allowing the 'freezing in' of metastable states without reaching the conditions required by the equilibrium phase diagram. Furthermore, the presence of ions (like H^+) which have a tendency to preferentially interact with sp^2 bonded C may be responsible for the removal of graphitically-bonded carbon in the growing film.

The properties of the films produced depend, among others, on the energy of the C ion impinging on the growing surface, as shown in Table. 9.1. Typical ion impact energies required for the growth of a–C:H with diamond-like properties are between 50 and 500 eV. These energies are above the sputter threshold for graphitic carbon, but below the energy where the sputter yield is greater than unity. This indicates that the ion flux, in addition to being the source of new material for growth, may also sputter away the newly deposited surface. Since the graphite bonds sputter at a greater rate than the diamond bonds, a matrix rich in sp^3 bonding results from this dynamic growth process. The presence of H^+ ions in the plasma also plays an important role in the film growth. H^+ ions are needed to saturate dangling bonds, and it is most likely that the H^+ ions are also instrumental in the removal of sp^2 bonded carbon by the formation of volatile radicals. Carbon and hydrogen ions with the required kinetic energy are usually obtained in a low pressure CVD system using either DC or RF generated plasmas of a hydrocarbon gas.

The properties of the as-grown films depend strongly on growth conditions such as gas mixture, substrate temperature and ion energy, as is demonstrated in the wide range of properties given in Table 2.2. Many of these properties depend on the amount of hydrogen incorporated in the films, which may vary from a few atomic percent up to $\sim 50\%$; typical films contain 30–40% of hydrogen. In this respect a–C:H differs from a–Si:H which contains only a few (~ 5) atomic percent of H, an amount which is sufficient to saturate all dangling bonds in Si. As is shown below, hydrogen loss from DLC films induced by ion bombardment is responsible for the changes in many of the properties of DLC films subjected to ion implantation.

The changes that occur in a–C:H films upon ion irradiation have been studied by several groups and the main results have been summarized in recent review articles [9.4, 5]. Hence there now exists a rather clear picture of the effects that the irradiation has on the material structure and composition. The most relevant parameters in these studies are, in close analogy to the case of ion-implantation in diamond, the ion mass and energy (both of which determine the dE/dx energy loss values), the ion fluence and the a–C:H film temperature during irradiation. Below, the currently available data will be briefly reviewed, concentrating on the results of the Technion group which has carried out an extensive set of measurements, employing most of the diagnostic techniques which have been employed to characterize ion-implanted diamond.

9.1.1 DC Conductivity

Resistivity changes as a result of ion irradiation are possibly the most spectacular and most intensively studied of all ion-induced effects. A wide variety of ions have been implanted into a–C:H and the resulting changes in conductivity were measured. Figure 9.1 shows the dose dependence of the resistivity of a–C:H films irradiated with C^+ (50 keV), Ar^+ (110 keV) and Xe^+ (270 keV) ions [9.6]. The ion energies were chosen so that the projected range in each case was about 1000Å. The experiments were carried out with the samples held at 100°C so as to avoid blistering which otherwise occurred for high dose noble gas implantations. The room temperature resistivity of the unirradiated films, ρ_0, was typically about 10^7 Ω cm.

The dose dependence of the resistivity is qualitatively the same for each ion studied and is amazingly similar to that found for an ion-implanted diamond crystal (Fig. 8.3 and note that $\rho = 1/\sigma$). Above a certain critical dose the resistivity drops (i.e., the conductivity rises) rapidly until it reaches an intermediate saturation level which appears as a knee in the curve. This knee seems to occur for all implantation conditions (ion species and energies) at an implantation dose at which the accumulated damage density (as calculated by TRIM) reaches a value of the order of the normal density of the material ($\sim 2\times10^{23}/cm^3$). This indicates that at this intermediate saturation level all atoms in the DLC target have on the average been dislodged once and most hydrogen atoms have been lost from the a–C:H material. Nevertheless the DLC has not yet completely graphitized at this stage. With still higher ion doses, a further decrease in ρ is observed [9.6], with the material gradually transforming into graphite.

Ingram and coworkers [9.7] have also measured the DC conductivity dependence on fluence in a similar geometry, but for 6.4 MeV fluorine ions and 1 MeV gold ions. Their irradiation parameters were chosen so that both ions deposit roughly the same amount of energy via electronic processes, yet the gold ions deposited some 350 times more energy to nuclear processes than the fluorine

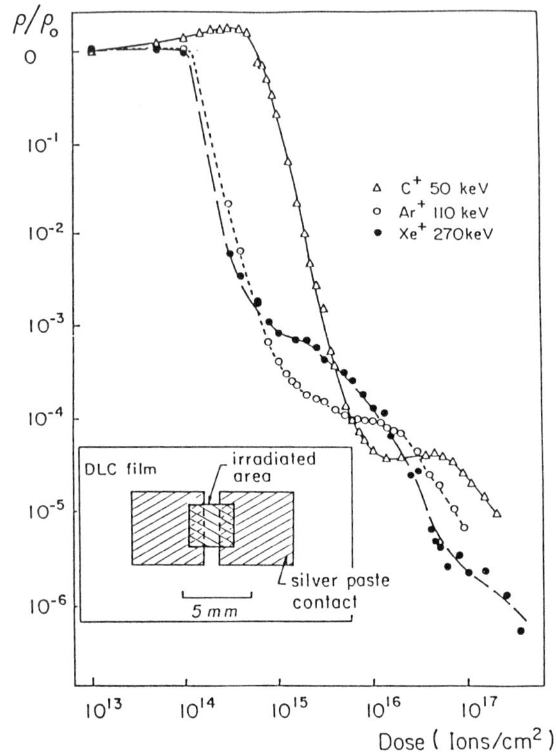

Fig. 9.1. Ion dose dependence of resistivity for C^+, Ar^+, and Xe^+ irradiation of an a-C:H film. *Inset*: the irradiation geometry [9.6]

ions. Their results, shown in Fig. 9.2, display a similar threshold behavior at the onset of the conductivity increase for both ions studied (which is similar to the results found in [9.6]). However the resistivity drop following this onset differs drastically for the two cases; the fluorine irradiation results in a drop in resistivity by some nine orders of magnitude while the resistivity changes for the gold irradiation are much less pronounced, leveling off after a reduction of only two or three orders of magnitude.

The large increase in conductivity measured as a function of irradiation dose invites close inspection; and, just like the case of diamond, study of the $\sigma(T)$ temperature dependence was directed toward increased understanding of the physics of the conduction mechanism. The temperature dependence of the conductivity for samples irradiated with different doses of 50 keV C^+ ions is shown in Figs. 9.3 and 9.4. In Fig. 9.3 $\ln\sigma(T)$ is displayed versus $1/T$ while in Fig. 9.4, the same data are displayed as $\ln\sigma(T)$ versus $T^{-1/4}$. It can be clearly seen that the data for the implanted films fit the functional dependence $\ln\sigma \propto T^{-1/4}$ (Fig. 9.4) better than $\ln\sigma \propto 1/T$ (Fig. 9.3). Hence it has been concluded in [9.6] that the mechanism responsible for the ion-induced conductivity in DLC

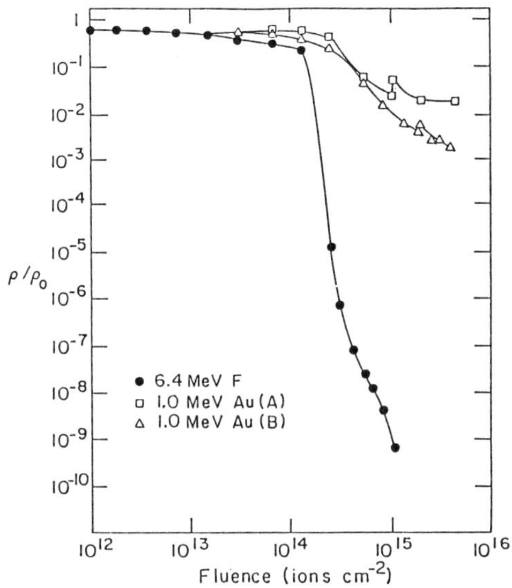

Fig. 9.2. Ion dose dependence of resistivity of DLC films for 6.4 MeV fluorine and 1 MeV and 10 MeV gold implants [9.8]

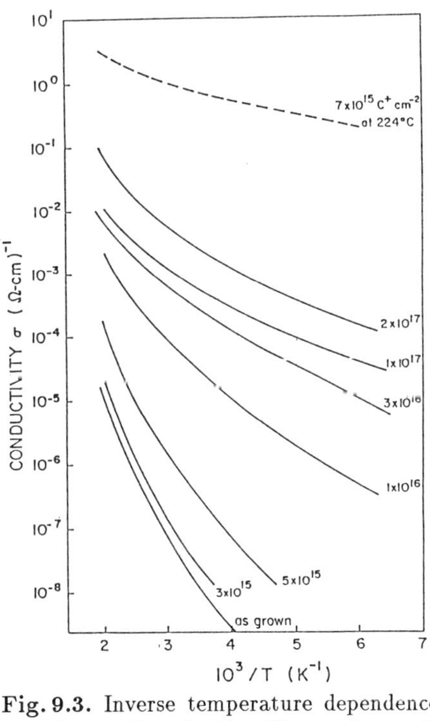

Fig. 9.3. Inverse temperature dependence of the conductivity (σ) for a–C:H film samples irradiated with different doses of C^+ ions. The broken line is for a sample irradiated with 7×10^{15} C^+ cm^{-2} at 224°C [9.6]

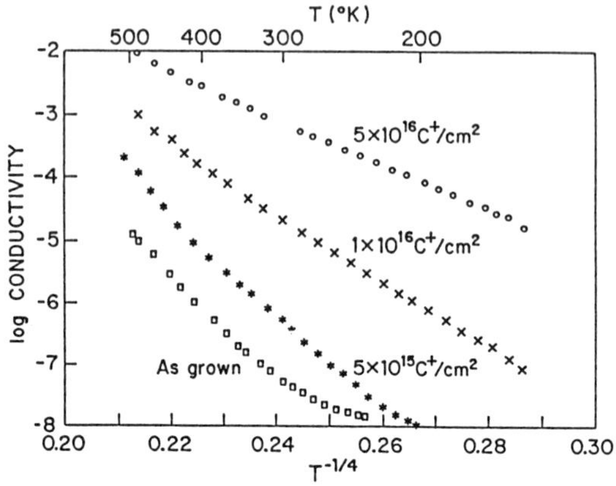

Fig. 9.4. Log of the conductivity versus $T^{-1/4}$ for the conductivity data shown in Fig. 9.3 for irradiated DLC films [9.6]

is, most likely, a variable range hopping mechanism, similar to that observed for ion-damaged diamond.

As for the determination of the carrier type responsible for the electrical conductivity, thermopower measurements [9.3] suggest that for lower dose implantations of C ions (70 keV, $5 \times 10^{15} cm^{-2}$) the conduction is via electron hopping, while for higher doses ($2 \times 10^{16} cm^{-2}$) it may be due to hole conduction.

9.1.2 Optical Characterization

Ion-implanted DLC films have been subjected to two kinds of optical measurements: absorption measurements from which the optical band-gap of the material has been deduced, and Raman spectroscopy which has yielded information on structural transformations that have been induced by the implantation.

As for the optical band gap, it has been found [9.6] that over a wide range of photon energies a linear relationship between photon energy (E) and $\sqrt{\alpha E}$ (α being the absorption coefficient) holds; hence Tauc plots can yield reliable information on the optical band gap of the material, as shown in Fig. 9.5. With increasing implantation dose, the optical gap shrinks, in agreement with the darkening in color which accompanies heavy dose implantations in DLC films.

The Raman spectra [9.9, 10] of unirradiated DLC in general exhibit a superposition of contributions from diamond-like structures, from crystalline graphite and from disordered graphite. The spectral lines, however, are greatly broadened due to the large variety of bonding configurations, the presence of disorder and the presence of hydrogen in these films. The Raman spectra emphasize the sp^2 over the sp^3 contribution because of its large Raman cross section, the contributions from the sp^3 diamond regions having a much smaller Raman cross section (Sect. 5.1). The Raman spectra and their lineshapes depend sensitively on the DLC film properties (Sect. 5.1), which in turn depend on

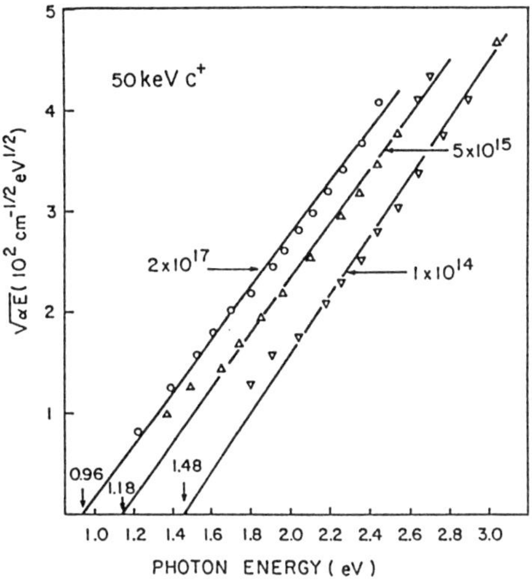

Fig. 9.5. Tauc plots of the photon energy (E) dependence of the optical absorption coefficient (α) plotted as $(\alpha E)^{1/2}$ on the ordinate for three different doses (ions/cm^2) of C^+ irradiation of a–C:H films at 50 keV [9.6]

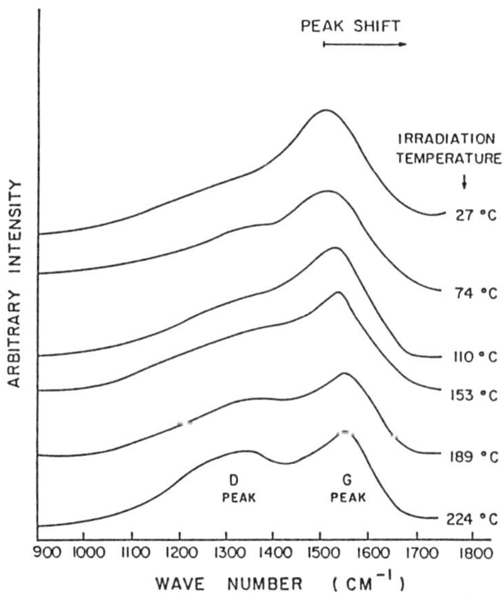

Fig. 9.6. Raman spectra for a–C:H irradiated with 50 keV C^+ ions to a dose of 7×10^{15} ions cm^{-2} at various substrate temperatures [9.6]

preparation technique. Upon successive irradiations, the Raman spectra for the DLC films show a shoulder at lower wave number, which can be associated with the disorder-induced D peak (Sect. 5.1). Its growth is greatly enhanced for elevated temperature implantations, as shown in Fig. 9.6, where the appearance

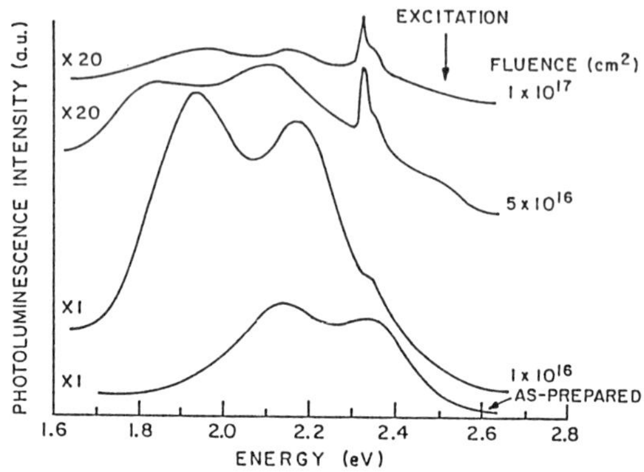

Fig. 9.7. Photoluminescence spectra for a–C:H irradiated with 100 keV H^+ ions to three different fluences [9.12]

of a well-resolved D peak is attributed to the formation of very small graphite crystallites at the highest substrate temperatures [9.6]. The G-peak normally associated with graphitic Raman-allowed vibrations is seen to shift to higher frequencies as the substrate temperature increases, again indicative of a decrease in the C–C distance as the tiny graphite crystallites form. Also of significance to the observation of Raman spectra in diamond and DLC films (Fig. 9.6) is the strain-induced shift of the Raman line due to the interaction of the film with various substrates and the loss of hydrogen in the case of ion-implanted DLC, both of which alter the in-plane C–C distances [9.11].

Photoluminescence (PL) studies by *Gonzales-Hernandez* and coworkers on H- and N-implanted DLC films [9.12] show that for relatively low fluence implantations ($1 \times 10^{16} H^+/cm^2$ at 100 keV) the PL intensity increases by a factor of 2, while for higher dose irradiations (5×10^{16}, $1 \times 10^{17} cm^{-2}$) the PL intensity decreases sharply, as shown in Fig. 9.7.

9.1.3 Structural Modifications and Hydrogen Loss

Two major structural modifications accompany ion-implantation of DLC and these may both be correlated to the other changes in the film properties described above. These are a loss of hydrogen and an increase in density of dangling bonds. The loss of hydrogen resulting from ion implantation in a–C:H film has been measured by several groups (Fig. 9.8) and has been modeled by *Adel* et al. [9.14]. The results of these studies indicate that for films with a typical H concentration of 30–40%, substantial H loss takes place at relatively low irradiation doses (10^{15} ions/cm^2). However, a saturation level of about 5 at.% H is reached, at which H loss practically stops. This behavior seems to be almost universal, and independent of irradiation conditions; MeV heavy ions, or sub-MeV light ions all exhibit a similar dependence for the H concentration on irradiation dose, regardless of the dominant physical mode of energy transfer from the

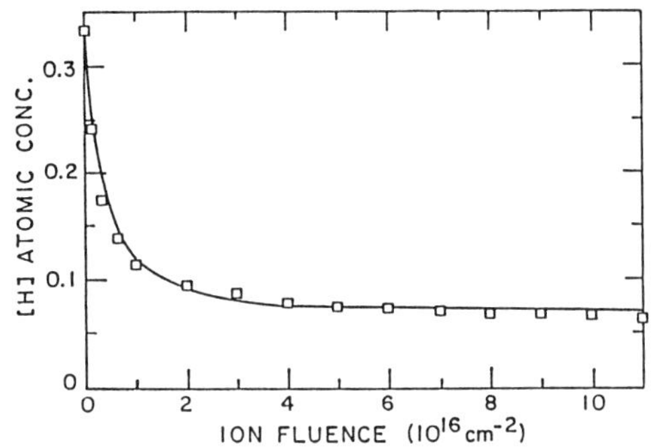

Fig. 9.8. Hydrogen atomic concentration in a–C:H as a function of irradiation dose with 6 MeV N^+ ions, according to *Baumann* et al. [9.13]. The *solid line* indicates the fit obtained by using the functional dependence from the bulk molecular recombination model of [9.14]

ion to the material. Whether the mechanism for H loss is nuclear or electronic stopping is not entirely clear; however evidence is given below suggesting that each may play a different role.

Baumann et al. [9.13] have fitted their measured hydrogen concentration versus dose curves with three different exponentials attributed to three different hydrogen release cross sections. These may be interpreted as the result of different C–H bonding configurations in the amorphous carbon matrix. *Sellschop* and coworkers [9.15] proposed a similar interpretation for their data on ion-beam-induced hydrogen release from natural diamond, suggesting that the changes in the slope in the logarithm of H concentration versus dose reflect different bonding states of H to C in diamond.

The mechanism by which H is lost from a–C:H films during thermal annealing has been illuminated in a series of experiments by *Wild* and *Koidl* [9.16]. By performing mass-resolved measurements of the species effused from sandwiches of deuterated and hydrogenated carbon films, they have shown that hydrogen always leaves the a–C:H films in molecular form, i.e., molecular recombination occurs in the bulk, followed by diffusion of the molecules through the layer until they are lost into the vacuum. A similar conclusion was also reached by *Moeller* who has shown that, during D^+ irradiation of H^+ implanted graphite, recombination of atomic to molecular hydrogen occurs in the bulk, and the effusing species are H_2 or HD molecules [9.17].

An alternative description of ion-irradiation induced hydrogen effusion from a–C:H films has recently been proposed by the Technion group [9.14]. It is based on the above described experimental evidence for molecular effusion and does not require the introduction of different C–H bonding configurations. The model is statistical in nature and is based on a second order kinetic process, namely the recombination of atomic to molecular hydrogen. The basic assumption in the model is that in order for a hydrogen molecule to be created in the bulk, two C–H bonds within some characteristic recombination length of each other

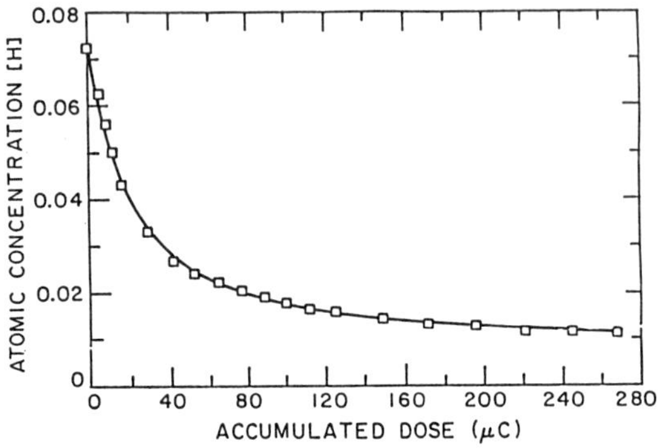

Fig. 9.9. Data for ion beam-induced (17 MeV ^{15}F ions) hydrogen release from a natural diamond (from [9.15]) fitted by the molecular recombination model of [9.14]

must be broken by the energy deposited through passage of the implanted ion. As a consequence of this requirement, the change in hydrogen concentration $d\rho$ due to a dose increase $d\phi$ is proportional to the number of hydrogen pairs in a characteristic volume V, so that

$$\frac{d\rho}{d\phi} = -K\rho(\rho V - 1) \tag{9.1}$$

where K is an effective molecular release cross-section. This differential equation has the solution

$$\rho(\phi) = [1/\rho_f + (1/\rho_0 - 1/\rho_f)\exp(-K\phi)]^{-1}. \tag{9.2}$$

Here ρ_0 is the initial volume hydrogen density and $V = 1/\rho_f$, which is the inverse of the final hydrogen concentration. The details of this calculation have been published elsewhere [9.14].

The model has been applied to all available published data on hydrogen effusion from a–C:H films using ρ_f and K as adjustable parameters. A typical example of the quality of the fits which can be obtained by this model is given in Fig. 9.8 which shows the atomic hydrogen concentration dependence on N^+ fluence, adapted from the data of *Baumann* et al. [9.13]. It is evident that the hydrogen concentration follows the functional dependence given by the model very closely; similar fits were obtained for all data analyzed, even for hydrogen release from natural diamond implanted with 17 MeV ^{15}F ions [9.15], as shown in Fig. 9.9. Table 9.2 summarizes the relevant experimental parameters for the various data including the release cross section K and the final hydrogen concentration ρ_f used to generate the fits. The physics governing the release mechanism resides in the parameter K and perhaps in the final H concentration ρ_f. The correlation between hydrogen loss during ion irradiation and the electronic stopping power of the incident ion has already been demonstrated by *Baumann* et al. [9.13]. The approximate correlation between K and the elec-

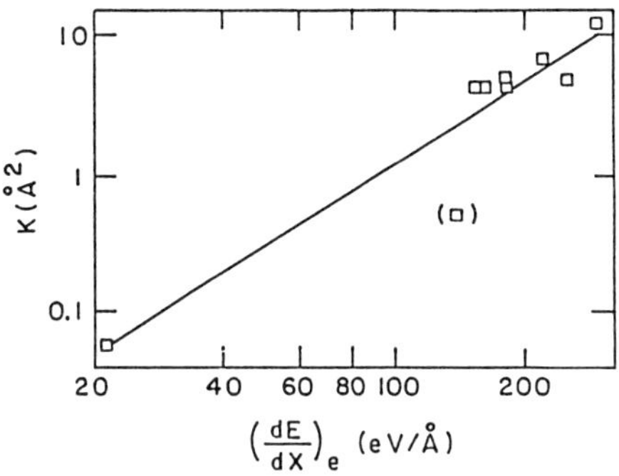

Fig. 9.10. Log-log plot of the cross section K for the molecular hydrogen release from a–C:H films versus electronic energy loss [$(dE/dx)_e$ as given in Table 9.2] [9.13]. The straight line has a slope of 2

Table 9.2. Values of parameters used to fit the bulk molecular recombination model for various experimental data on ion-induced hydrogen release from a–C:H. The electronic and nuclear energy loss for each case is also listed

Ion	Energy [Mev]	K Cross Section [Å^2]	$(dE/dx)_e$ electronic [eV/Å]	$(dE/dx)_n$ nuclear [eV/Å]	ρ_f [at. %]	Reference
^{15}N	7	0.5±0.03	140	1.0	6.5	[9.13]
^{15}N	3.5	4.5±0.5	155	1.4	10	[9.13]
^{20}Ne	1.6	4.5±0.5	160	6	9	[9.13]
^{40}Ar	2.15	4.5±0.5	180	22	10	[9.13]
^{58}Ni	4	13±3	192	17	7	[9.18]
^{19}F	6.4	5	250	0.67	6	[9.8]
^{197}Au	1	7	221	307	20	[9.8]
^{12}C	12	5	100	0.14	9	[9.19]
^{12}C	0.05	0.06±0.02	15–27	18–14	5	[9.6]

tronic stopping powers listed in Table 9.2 is illustrated in Fig. 9.10 where the effective cross section K is plotted versus the energy loss due to electronic processes. Although the points are somewhat scattered, a least squares fit to this data produced a line of slope 2.0 with standard deviation 0.3. This is in fact the value expected for a molecular release cross section governed by the probability of independently freeing two hydrogen atoms within a characteristic recombination distance.

The abnormally small decrease in resistivity following Au implantation, observed by *Ingram* and coworkers [9.8] and shown in Fig. 9.2, can well be explained by the above model. The very dense damage cascade, which accompanies the slowing down of the heavy gold particles has a high density of broken

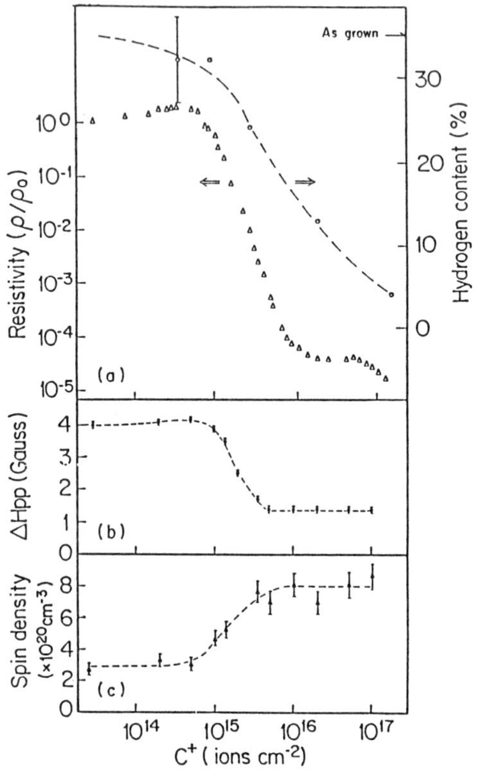

Fig. 9.11. Variation with 50 keV C^+ ion dose of (**a**) the resistivity and hydrogen content, (**b**) the ESR resonant linewidth and (**c**) the spin density of a–C:H films [9.20]

bonds, all of which compete with the liberated hydrogen atoms for the capture of hydrogen, the former to saturate the bonds, and the latter for H_2 molecule formation. Due to this competition, fewer H_2 molecules are formed, the films remain H-rich (a point which should be verified experimentally!), and hence their resistivity remains high.

Since hydrogen plays an important role in stabilizing the a–C:H structures, hydrogen loss should be accompanied by bond breakage. The bond breakage may in turn lead to the creation of unsaturated dangling bonds that can be observed by electron spin resonance (ESR) measurements. Such measurements can, as mentioned in Sect. 5.6, provide the total number of unpaired spins in the sample under study and, if the affected volume (i.e., the thickness of the damaged layer) is known, the average spin density can be estimated. Furthermore, from the ESR line-shape and resonant field, information on spin-spin interactions and on the C–C binding can be deduced. The data of *Adel* et al. [9.20] for 50 keV C implanted a–C:H films (Fig. 9.11) show that below a critical dose of $\sim 7\times10^{14}$ cm^{-2} no changes in the magnitude or line-shape of the ESR signal can be detected, indicating that within the sensitivity limit of ESR, no new dangling bonds are formed by the implantation process. However, for doses in the range $7 \times 10^{14} < \phi < 1 \times 10^{16}$ cm^{-2}, an increase in the spin density by about a factor of ~3.5 is observed, accompanied by a narrowing of the resonance linewidth; see panels (b) and (c) in Fig. 9.11. The increase in ESR signal shows that bonds are now broken by the implantation, but their number is far smaller than the number of implanted ions. The narrowing of the line width is indicative of the spatial proximity of these spins. For doses exceeding 10^{16}cm^{-2}, the dangling bond density saturates at the very high value of 8×10^{20} cm^{-2}. It should be noted that no shift in resonance position accompanies these changes at any stage, indicating that no graphitization has taken place even for the highest doses studied.

9.1.4 Attempts to Dope a–C:H by Ion-Implantation

Some attempts have been made to dope a–C:H films electrically by implanting acceptor and donor ions. *Wong* and coworkers [9.21] have, most recently, implanted a–C:H films by B^+, BF_2^+, and P^+. These implantations were found to increase the conductivity of the film, once a certain threshold dose had been reached. This increase has been attributed to an electrical doping effect [9.21]. Similar experiments have been carried out by *Amir* and *Kalish* [9.22, 23], in which diamond-like films have been implanted with $^{11}B^+$ and, as a control under identical conditions, also with $^{12}C^+$ ions (both leading to similar damage due to the proximity in projectile mass). These measurements have not revealed any significant difference between the C and B implantations, despite the valence difference between carbon (the host material) and boron (a potential acceptor). It is therefore concluded in this work that the observed electrical effect was due to implantation-induced damage of the kind discussed above, and not to chemical doping effects caused by the particular implanted ions.

Cheng et al. [9.24] have attempted to electrically dope a–C:H films p type by low energy (5 keV) high dose (0.5–20.0×10^{16}/cm^2) boron implantation. A decrease in film resistivity of 2–7 orders of magnitude was observed to result from the B implantation which could be further enhanced by annealing at 500°C for 4 hours. Control experiments of C implantations at high doses ($> 10^{17}$ cm^{-2}) performed under conditions similar to those of the B implantations, have also led to a rapid drop in resistivity. However, the onset of this drop was found to occur at C doses somewhat higher than for the B implantations, a fact which was taken as proof that boron behaves as a dopant when implanted into DLC. No studies on the physical, structural or chemical state of the implantation-affected layer have however been reported.

When implantations are carried out into heated DLC (e.g., $T_i \sim 200$°C) all effects described above are enhanced, and thus can be observed at lower ion doses [9.6]. In particular, complete graphitization, as observed in the Raman spectra as well as in the mechanical and chemical properties of the irradiated films, already take place at moderate doses. This enhancement should probably be traced back to the enhanced diffusion processes (H out-diffusion and vacancy and interstitial diffusion), all of which exhibit a sharp, exponential rise with temperature.

9.1.5 Discussion of Implantation-Induced Effects in DLC

Amorphous hydrogenated carbon films have some features which make them "diamond-like" and others which make them very different from crystalline diamond. The diamond features of the films are governed by the sp^3 bonding which gives the material its hardness, transparency and electrical insulating properties (Table 2.2). The major differences between DLC and crystalline diamonds are associated with the amorphous features of the material and with its high hydrogen content. The structure of the a–C:H has been described in terms of a two-phase model along the lines suggested by *McKenzie* and coworkers [9.25], as well as *Robertson* and *O'Reilly* [9.26]. In this model, hard a–C:H material is considered to consist of sp^2 clusters, typically planar aromatic ring structures, which are interconnected by randomly oriented tetrahedral sp^3 bonds. The hydrogen atoms may be bonded on either the tetrahedral sites, where the hydrogens are required to reduce bond angle disorder, or on the edges of the ring structures where they cause terminations.

Based on this model, it is clear that significant loss of hydrogen must lead to major structural changes. Firstly, since a random tetrahedral phase is highly over-constrained (analogous to a–Si:H), and since the hydrogen is responsible for reducing the co-ordination number, its removal is expected to destabilize the sp^3 phase. This would either increase the number of unsatisfied carbon bonds or result in a restructuring whereby two previously sp^3 C–H bonds "recombine" to form an sp^2 bond. The latter effect would result in a net conversion from sp^3 type bonding to sp^2 type bonding. The growth of the sp^2 clusters would be further enhanced by the removal of hydrogen from the ring edges, so that a net increase in cluster size is expected. The fact that ion bombardment breaks bonds and thus leads to hydrogen release, as described above, suggests that growth of sp^2 (graphite-like) clusters will be found in damaged a–C:H. Thus it is expected that the major changes in the electrical and optical properties of ion-implanted DLC should go hand-in-hand, and both types of changes should set in at about the same damage levels (i.e., the same ion dose for a given implantation condition). This indeed is the case, as can be seen in Fig. 9.11. For 50 keV C^+ implantations, hydrogen loss starts at the same dose at which changes in the ESR signal width and magnitude become noticeable. At the same dose regime where these structural changes occur, other film properties also undergo dramatic changes — the electrical resistivity drops sharply (Fig. 9.11a) and the optical band gap shrinks (not shown) [9.6]. It is interesting to note the similarity in the ion-beam induced electrical conductivity between an implanted diamond crystal (Fig. 8.3) and implanted a–C:H films, despite the basic differences that exist in their structure. One possible reason for this similarity, which requires further investigation, is that part of the onset of the observed conductivity has to do with the percolative formation of a conductive pathway via the implanted ions (whether carbon or metals), and is not to be related with bond breakage or hydrogen loss. This point was recently investigated by *Prawer* and *Kalish* [9.27] who implanted diamond and quartz with 320 keV Xe or 50 keV C ions at ever increasing doses while measuring the changes in electrical conductivity. They have observed strong increases in conductivity for the case of quartz implanted with C at about the same critical dose as was found for similarly implanted diamond and a–C:H films, while no change in conductivity could be detected for Xe implanted quartz. This result indicates that at least part of the increase in electrical conductivity measured for ion-implanted diamond should not be related just to the formation of buried graphitic conductive pathways (which cannot occur in quartz) but is due to the conductivity of the implanted species (carbon) itself.

More details on the ion-beam-induced modifications of a–C:H films, on the theoretical model regarding the structural changes that the material undergoes, and on the electrical conduction mechanism in these films can be found in recent review papers [9.4, 5].

9.2 Diamond Films

As already mentioned in Sect. 2.1.7, diamond films, as grown most recently by a wide variety of CVD methods, differ from single-crystal diamonds in that they are polycrystalline in nature with many grain boundaries and defects, and often contain some graphitic or amorphous carbon impurities. These differences result in important deviations from those of the natural single crystals with regard to the physical, chemical, and optical properties. For example, the films grain boundaries and defects are thought to be responsible for the unusual positive temperature coefficient of the thermal conductivity of CVD films in the temperature range 300-700 K in contrast to the known negative slope typical of single-crystal diamond. It is therefore by no means *a priori* clear that the effects of ion beams on diamond films are necessarily expected to be the same as for single crystal diamond. In fact, one might expect grain boundaries, graphitic impurities, and defects to have an important influence on the electrical behavior of ion-beam-irradiated diamond films.

An experiment designed to compare the changes in the electrical conductivity in ion-implanted CVD diamond films to those induced in identically implanted diamond single crystals has recently been carried out [9.27]. In that work, films consisting of 3-6μm diamond crystallites (grown on quartz) have been placed next to type IIa diamond slabs, and both have been subjected to 100 keV C or 320 keV Xe implantations at 20 or 200^{0}C. The electrical resistivities of the specimen pairs have been measured in situ, as a function of ion dose, in a way similar to that previously employed to study diamond-like carbon (DLC) films [9.14] (Fig. 9.1) and diamond crystals [9.28, 29] (Figs. 8.1 and 8.3). The results of this comparative study are shown in Fig. 9.12. Interestingly,

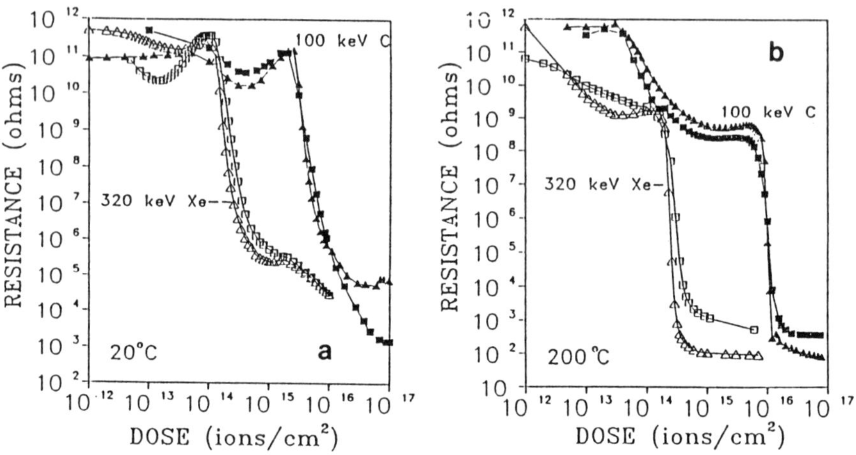

Fig. 9.12. Resistance vs dose for irradiations carried out at (**a**) 20±2°C and (**b**) 200±10°C for CVD diamond films (*squares*) and natural type IIa single-crystal diamond (*triangles*). The open symbols are for irradiation with 320 keV Xe, while the closed symbols refer to irradiation with 100 keV C ions (from [9.2])

the changes in resistivity obtained for the implanted diamond films are remarkably similar to those obtained for the identically implanted diamond crystals for all cases studied. (And they also resemble, in their general trend, the dependence of the resistivity on ion dose for ion implanted DLC.) This similarity in the response to ion implantation of polycrystalline and single-crystal diamond suggests that the conduction mechanism in ion-beam-irradiated polycrystalline CVD diamond films is not dominated by grain boundaries and graphitic impurities, as one might have expected, but rather is determined by the intrinsic properties of diamond itself.

10. Concluding Remarks

The carbon-based materials and their response to ion-implantation and thermal treatments, which have been reviewed here, are unique in many aspects. The basic carbon materials considered, diamond and graphite, have many differences. Diamond exhibits 3-dimensional sp^3 tight tetrahedral bonding, and unusual physical properties result from this bonding (hardness, high thermal conductivity, optical transparency, high index of refraction and wide band-gap). Graphite exhibits a sp^2 bonded 2-dimensional layered structure which leads to highly anisotropic nearly metallic properties. These differences in bonding make the properties of these two allotropic forms of carbon about as different as one can imagine. Yet, as shown in the preceding chapters, ion implantation can cause non-reversible changes from diamond to graphite under conditions which are now fairly well understood.

Between the two extreme forms of carbon (diamond and graphite), many intermediate, interesting and useful carbon-based materials can be found, e.g., the "so called fullerenes" [10.1]. These range from amorphous diamond-like carbon (with or without hydrogen) to amorphous graphite, from the tangled network of one-dimensional polymers to carbon fibers. Obviously not all of the carbon-based materials have yet been subjected to ion implantation, and much work along the lines outlined above still needs to be done. Interestingly, it seems as if even the diamond system which has been studied now for over 20 years still lacks many investigations of the kind recently applied to the more novel materials like a–C:H. It is therefore hoped that this review, together with the renewed interest in diamond, triggered by the possibility of growing diamond films, will lead to revisiting many important experiments described above, using the additional know how and the improved techniques available today. Furthermore, other carbon-based materials should also be subjected to ion implantation and to measurements similar to those carried out on the more basic systems, thus hopefully finding technologically important modifications of these materials.

Despite the very basic differences between the sp^2 bonded and the sp^3 bonded carbon materials, they exhibit some common features regarding their response to ion-implantation, and these common features may be unique to all the carbon materials. These have to do with the fact that the target material consists of one of the lightest elements ($M_2 = 12$); hence most of the implanted ions (M_1) will be more massive than the target atom ($M_1 > M_2$), a condition which affects the momentum transfer during the collision. Therefore, both implanted graphite and implanted diamond exhibit unusual swelling, much more than can be accounted for by just the additional atoms forced into the matrix. This swelling has been attributed to the spatial separation between vacancies and recoil interstitials [10.2]. The resulting local density imbalance has important implications on the ability of the disrupted system to regrow to its original structure upon annealing. It is interesting to note here that, in contrast to experiments performed on the annealing of implanted diamond, where excess vacancies are claimed to inhibit diamond regrowth [10.2], recent calculations [10.3] show that large vacancy concentrations change the relative binding energies so that diamond becomes more stable than graphite, thus permitting nucleation and growth of diamond films.

Other compositional changes which have been observed to be caused by ion implantation in hydrogen-containing carbon-based materials are hydrogen loss. This, being a non-reversible process, leads to the transformation of the bombarded material (a–C:H or polymers for example) to a new form, which is usually more graphite-like. Interestingly, it has been shown that the hydrogen loss and the resulting transformation to hydrogen-deficient materials is not solely caused by the ion impact. Other kinds of focused energy deposition, such as electron [10.4] or photon irradiation [10.5], also lead to the transformation of a–C:H, closely resembling those caused by ion impact. This field, of exposing carbon-based materials to other forms of energy deposition, has not been much explored so far. In this context it could be particularly interesting, for example, to shine high energy photons onto diamond, graphite or related materials held under high pressures to investigate possible graphite to diamond transformation.

The technological implications of ion implantation into carbon-based materials have hardly been addressed so far. The finding that both glassy carbon and diamond can be further hardened by ion-implantation is intriguing and may find industrial uses. The doping of diamond p- and n-type and the realization of electronic or opto-electronic devices based on the semiconducting properties of diamond are most tempting. Much work is therefore currently devoted to this topic both by ion-implantation and by epitaxial growth of doped diamond layers. There is still much need for fundamental theoretical work [10.6] on the stability of diamond, on diffusion mechanisms in diamond, and on potential dopants for this interesting semiconductor. The fact that diamond and diamond-like materials can be locally turned graphite-like by exposing selected areas to ion implantation or other forms of focused energy deposition, may be important in producing conductive or opaque pathways in diamond or in related transparent insulating materials. And finally, the understanding of the interaction of carbon materials with energetic charged particles is of great importance as these materials may find important uses in space, or in fusion reactor technology, where they are bound to be subjected to ion-induced radiation damage.

We hope that the present review will trigger research both on the basic and on the applied levels to a family of materials of great technological importance and of much interest to materials science and to the life sciences.

References

Chapter 2

2.1 F.P. Bundy: J. Geophys. Res. **85**, 6930 (1980)
2.2 F.P. Bundy: In *Solid State Physics under Pressure: Recent Advance with Anvil Devices*, ed. by S. Minomura (Reidel, Dordrecht 1985) p.1
2.3 B.T. Kelly: *Physics of Graphite* (Applied Science, London 1981)
2.4 C. Kittel: *Introduction to Solid State Physics*, 6th edn. (Wiley, New York 1986)
2.5 M.S. Dresselhaus, J. Steinbeck: Tanso **132**, 44 (1988)
2.6 J. Krumhansl, H. Brooks: J. Chem. Phys. **21**, 1663 (1953)
2.7 J. Robertson: Adv. Phys. **35**, 317 (1986)
2.8 S. Prawer, C.J. Rossouw: J. Appl. Phys. **63**, 4435 (1988)
2.9 Y. Bar-Yam, T.D. Moustakas: Nature **342**, 786 (1989)
2.10 R.W.G. Wyckoff: *Crystal Structures* (Interscience, New York 1964)
2.11 M.S. Dresselhaus, G. Dresselhaus, K. Sugihara, I.L. Spain, H.A. Goldberg: *Graphite Fibers and Filaments*, Springer Ser. Mater. Sci., Vol.5 (Springer, Berlin, Heidelberg 1988)
2.12 I.L. Spain: In *Chemistry and Physics of Carbon*, ed. by P.L. Walker Jr., P.A. Thrower (Dekker, New York 1981) p.119
2.13 S.B. Austerman: In *Chemistry and Physics of Carbon*, ed. by P.L. Walker Jr. (Dekker, New York 1968) p.160
2.14 Y. Kaburagi: Priv. communication (1991)
2.15 A.W. Moore: In *Chemistry and Physics of Carbon*, ed. by P.L. Walker Jr., P.A. Thrower (Dekker, New York 1973) p.69
2.16 K. Matsubara, T. Tsuzuku, M. Murakami: Extended Abstracts, 20th Biennial Conf. on Carbon (1991) p.556
2.17 H. Yasujima, M. Murakami, S. Yoshimura: Appl. Phys. Lett. **49**, 499 (1986)
2.18 Y. Hishiyama, S. Yasuda, M. Inagaki: J. Mater. Sci. **23**, 3722 (1988)
2.19 M. Inagaki, S. Harada, T. Sato, T. Nakajima, Y. Horino, K. Morita: Carbon **27**, 253 (1988)
2.20 Y. Hishiyama, Y. Kaburagi, M. Inagaki: In *Chemistry and Physics of Carbon*, ed. by P.A. Thrower (Dekker, New York 1991) p.1
2.21 L. Salamanca-Riba, G. Braunstein, M.S. Dresselhaus, J.M. Gibson, M. Endo: Nucl. Instr. Meth. Phys. Res. B **7** & **8**, 487 (1985)
2.22 G.M. Jenkins, K. Kawamura: *Polymeric Carbons - Carbon Fibre, Glass and Char* (Cambridge Univ. Press, London 1976)
2.23 R.R. Saxena, R.H. Bragg: J. Non-Cryst. Solids **28**, 45 (1978)
2.24 F. Rousseaux, D. Tchoubar: Carbon **15**, 55, 63 (1977)
2.25 A. Oberlin: In *Chemistry and Physics of Carbon*, ed. by P.A. Thrower (Dekker, New York 1989) p.1
2.26 A. Yoshida, Y. Kaburagi, Y. Hishiyama: Carbon **29**, 1107 (1991)
2.27 Y. Shiraishi: *Introduction to Carbon Materials* (Carbon Society of Japan, Tokyo 1984) pp.29-40 [Kaitei Tansozairyo Nyumon (in Japanese)]
2.28 Y. Kaburagi, S. Yasuda, Y. Hishiyama: Extended Abstracts, 18th Biennial Conf. on Carbon (1987) p.476

2.29 Y. Kaburagi, Y. Hishiyama, D.F. Baker, R.H. Bragg: Phil. Mag. B **54**, 381 (1986)
2.30 K. Ueno, T. Kumihashi, K. Saiki, A. Koma: Jpn. J. Appl. Phys. Lett. **27**, L759 (1988)
2.31 M.S. Dresselhaus, G. Dresselhaus: Adv. Phys. **30**, 139 (1981)
2.32 H. Zabel, S.A. Solin (eds.): *Graphite Intercalation Compounds I: Structure and Dynamics*, Springer Ser. Mater. Sci., Vol.14 (Springer, Berlin, Heidelberg 1990)
2.33 W. Rüdorff, E. Shultze: Z. anorg. allg. Chem. **277**, 156 (1954)
2.34 F.P. Bundy, J.S. Kasper: J. Chem. Phys. **46**, 3437 (1967)
2.35 R.C DeVries: Ann. Rev. Mater. Sci. **17**, 161 (1987)
2.36 J.C. Angus, C.C. Hayman: Science **241**, 877 (1988)
2.37 J.C. Angus, C.C. Hayman: Science **241**, 913 (1988)
2.38 W.A. Yarbrough, R. Messier: Science **247**, 688 (1990)
2.39 B.V. Spitsyn, L.L. Bouitov, B.V. Derjaguin: J. Cryst. Growth **52**, 219 (1981)
2.40 J. Narayan: J. Mater. Res. **5**, 2414 (1990)
2.41 S. Prawer, A. Hoffman, R. Kalish: Appl. Phys. Lett. **57**, 2187 (1990)
2.42 A. Feldman, H.P.R. Frederikse, X.T. Ying: Proc. SPIE **1146**, 78 (1990)
2.43 M.W. Geis: In *Diamond, Boron Nitride, Silicon Carbide and Related Wide Bandgap Semiconductors*, ed. by J.T. Glass, R.F. Messier, N. Fujimori. Proc. MRS **162**, 15 (1990)
2.44 J. Narayan, V.P. Godbole, C-W. White: Science **252**, 416 (1991)
2.45 P. Koidl, P. Oelhafen (eds.): *Amorphous Hydrogenated Carbon Films*, Proc. E-MRS **17** (Les Editions de Physique, Paris 1987)
2.46 G.H. Johnson, A.R. Badzian, M.W. Geis (eds.): *Diamond and Diamond-Like Materials Synthesis* (MRS, Pittsburgh, PA 1988)
2.47 D.R. McKenzie, R.C. McPhedram, N. Savvides, D.J.H. Cockayne: Thin Solid Films **108**, 247 (1983)
2.48 J. Robertson, E.P. O'Rielly: Phys. Rev. B**35**, 2946 (1987)
2.49 R. Kalish, E. Adel: In *Properties and Preparation of Amorphous Carbon Films*, ed. by J.J. Pouch, S.A. Alterowitz, Mater. Sci. Forum, Vols. 52 & 53 (Trans Tech., Aedermannsdorf, Switzerland 1990) p.427
2.50 Y. Lifshitz, S.R. Kasi, J.W. Rabalais: Materials Science Forum **52&53**, 237 (1990) "Properties and characterization of amorphous carbon films", ed. by J.J. Pouch, S.A. Alterovitz, published by the Solid State Institute, Technion, Israel
2.51 M.S. Dresselhaus, G. Dresselhaus: In *Light Scattering in Solids III*, ed. by M. Cardona, G. Güntherodt, Topics Appl. Phys., Vol.51 (Springer, Berlin, Heidelberg 1982) Chap.2
2.52 R. Al-Jishi, G. Dresselhaus: Phys. Rev. B **26**, 4514 (1982)
2.53 A. Zunger: Phys. Rev. B **17**, 626 (1978)
2.54 J.W. McClure: In *Proc. Int'l Conf. on Semimetals and Narrow Gap Semiconductors*, ed. by D.L. Carter, R.T. Bate (Pergamon, New York 1971) p.127
2.55 I.L. Spain: In *Chemistry and Physics of Carbon*, ed. by P.L. Walker Jr., P.A. Thrower (Dekker, New York 1973) p.1
2.56 V. Bayot, L. Piraux, J.P. Michenaud, J.P. Issi, M. Lelaurain, A. Moore: Phys. Rev. B **41**, 11770 (1990)
2.57 D.F. Baker, R.H. Bragg: Phys. Rev. B **28**, 2219 (1983)
2.58 H.R. Philipp: Phys. Rev. B **16**, 2896 (1977)
2.59 E.A. Taft, H.R. Philipp: Phys. Rev. **138**, A197 (1965)
2.60 R. Bacon: J. Appl. Phys. **31**, 283 (1960)
2.61 J.E. Field: In *Properties of Diamond*, ed. by J.E. Field (Academic, New York 1979) p.281
2.62 F. Herman, R.L. Kortum, C.D. Kuglin, R.A. Short: J. Phys. Soc. Jpn. (Suppl.) **21**, 7 (1966)
2.63 W. Saslow, T.K. Bergstresser, M.L. Cohen: Phys. Rev. Lett. **16**, 354 (1966)

2.64 C.D. Clark, E.W.J. Mitchell, B.J. Parsons: In *The Properties of Diamond*, ed. by J.E. Field (Academic, New York 1979) p.23

2.65 Y.S. Touloukian, R.W. Powell, C.Y. Ho, P.G. Klemens (eds.): In *Thermophysical Properties of Matter*, Vol.2 (Plenum, New York 1970) diamond, p.10; pyrolytic graphite, p.32

2.66 T.R. Anthony, W.F. Banholzer, J.F. Fleischer, L.-H. Wei, P.K. Kuo, R.L. Thomas, R.W. Pryor: Phys. Rev. B **42**, 1104 (1990)

Chapter 3

3.1 W.K. Chu, J.W. Mayer, M.A. Nicolet: *Backscattering Spectrometry* (Academic, New York 1978)

3.2 L.C Feldman, J.W. Mayer, S.T. Picraux: *Materials Analysis by Ion Channeling* (Academic, New York 1982)

3.3 J. Lindhard, M. Scharff, H.E. Schiøtt: Mat. Fys. Medd. Dan. Vid. Selsk. **33**, 14 (1963)

3.4 W. Eckstein: *Computer Simulation of Ion-Solid Interactions*, Springer Ser. Mater. Sci., Vol.10 (Springer, Berlin, Heidelberg 1991)

3.5 J. Ziegler, J.P. Biersack, U. Littmark: In *The Stopping Range of Ions in Solids* (Pergamon, New York 1985)

3.6 J. Ziegler, J.P. Biersack, J. Cuomo: Unpublished (1980)

3.7 J.P. Biersack, L.G. Haggermark: Nucl. Instr. Meth. Phys. Res. **174**, 257 (1980)

Chapter 4

4.1 W.K. Chu, J.W. Mayer, M.A. Nicolet: *Backscattering Spectrometry* (Academic, New York 1978)

4.2 L.C. Feldman, J.W. Mayer, S.T. Picraux: *Materials Analysis by Ion Channeling* (Academic, New York 1982)

4.3 D.V. Morgan: *Channeling* (Wiley, New York 1973)

4.4 L.C. Feldman, J.W. Mayer: *Fundamentals of Surface and Thin Film Analysis* (North-Holland, New York 1986)

4.5 W. Eckstein: *Computer Simulation of Ion-Solid Interactions*, Springer Ser. Mater. Sci., Vol.10 (Springer, Berlin, Heidelberg 1991)

4.6 L.C. Feldman, S.T. Picraux: In *Ion Beam Handbook for Materials Analysis*, ed. by J. W. Mayer, E. Rimini (Academic, New York 1977)

4.7 S. Connell, K. Bharuth-Ram, J.P.F. Sellschop, M. C. Stemmet, H. Appel: Nucl. Instr. Meth. Phys. Res. **B35**, 423 (1988)

4.8 J.P.F. Sellschop: In *Properties of Diamond*, ed. by J.E. Field (Academic, London 1979) p.107

4.9 R.A. Spits, T.E. Derry, J.F. Prins, J.P.F. Sellschop: Nucl. Instr. Meth. Phys. Res. B **51**, 63 (1990)

4.10 H. Baumann, T. Rupp, K. Bethge, P. Koidl, C. Wild: In *Amorphous Hydrogenated Carbon Films*, ed. by P. Koidl, P. Oelhafen (Les Editions de Physique, Paris 1987) p.149

4.11 J.P.F. Sellschop, C.C.P. Madiba, H.J. Annegarn: Nucl. Instr. Meth. Phys. Res. **168**, 529 (1980)

4.12 C. Wild, P. Koidl: Appl. Phys. Lett. **51** 19 (1987)

4.13 W. Moeller, P. Borgesen, B.M.U. Scherzer: Nucl. Instr. Meth. Phys. Res. B **19&20**, 826 (1987)

4.14 D.C Ingram, A. W. McCormick: Nucl. Instr. Meth. Phys. Res. B **34**, 68 (1988)

4.15 W. Brandt: Sci. Am. **218**, 90 (March 1968)

4.16 D. Boutard, W. Müller, B.M.U. Scherzer: J. Appl. Phys. **67**, 163 (1990)
4.17 A.W. Moore: In *Chemistry and Physics of Carbon*, ed. by P.L. Walker Jr., P.A. Thrower (Dekker, New York 1973) p.69
4.18 S.B. Austerman: In *Chemistry and Physics of Carbon*, ed. by P.L. Walker Jr (Dekker, New York 1968) Vol.4, p.160
4.19 A.W. Moore: In *Chemistry and Physics of Carbon*, ed. by P.L. Walker Jr., P.A. Thrower (Dekker, New York 1981) p.233
4.20 G. Timp, M.S. Dresselhaus: J. Phys. C **17**, 2641 (1984)
4.21 D. Sigurd, R.W. Sower, W.F. van der Weg, J.W. Mayer: Thin Solid Films **19**, 319 (1973)
4.22 H. Ishiwara, S. Furukawa: J. Appl. Phys. **47**, 1686 (1970)
4.23 B.S. Elman, G. Braunstein, M.S. Dresselhaus, G. Dresselhaus, T. Venkatesan, B. Wilkens: J. Appl. Phys. **56**, 2114 (1984)
4.24 T. Iwata, K.I. Komaki, H. Tomimitsu, K. Kawatsura, K. Ozawa, K. Doi: Rad. Eff. **24**, 63 (1975)
4.25 J.H. Barrett: Phys. Rev. B **3**, 1527 (1971)
4.26 D.S. Gemmell, R.C. Mikkelson: Phys. Rev. B **6**, 1613 (1972)
4.27 S.T. Picraux, J.A. Davies, L. Eriksson, N.G.E. Johansson, J.W. Mayer: Phys. Rev. **180**, 873 (1969)
4.28 L. Salamanca-Riba, G. Timp, L.W. Hobbs, M.S. Dresselhaus: In *Proc. Symp. on Intercalated Graphite*, ed. by M.S. Dresselhaus, G. Dresselhaus, J.E. Fischer, M.J. Moran (North-Holland, New York 1983) p.9
4.29 P. Boolchand, W.J. Bresser, D. McDaniel, K. Sisson, V. Yeh, P.C. Eklund: Solid State Commun. **40**, 1049 (1981)
4.30 B.S. Elman, L. Salamanca-Riba, M.S. Dresselhaus, T. Venkatesan: J. Appl. Phys. **55**, 894 (1984)
4.31 L. Salamanca-Riba, B.S. Elman, M.S. Dresselhaus, T. Venkatesan: In *MRS Symp. on Ion Implantation and Ion Beam Processing of Materials*, ed. by G.K. Hubler, O.W. Holland, C.R. Clayton, C.W. White (Elsevier, New York 1984) p.481
4.32 M.S. Dresselhaus, G. Dresselhaus: Adv. Phys. **30**, 139 (1981)

Chapter 5

5.1 R.O. Dillon, J.A. Woollam, V. Katkanant: Phys. Rev. B **29**, 3482 (1984)
5.2 R.P. Vidano, D.B. Fishbach: J. Am. Ceram. Soc. **61**, 13 (1978)
5.3 N. Wada, P.J. Gaczi, S.A. Solin: J. Non-Cryst. Solids **35-36**, 543 (1980)
5.4 F. Tuinstra, J.L. Koenig: J. Chem. Phys. **53**, 1126 (1970)
5.5 B.S. Elman, M.S. Dresselhaus, G. Dresselhaus, E.W. Maby, H. Mazurek: Phys. Rev. B **24**, 1027 (1981)
5.6 B.S. Elman, M. Shayegan, M.S. Dresselhaus, H. Mazurek, G. Dresselhaus: Phys. Rev. B **25**, 4142 (1982)
5.7 M.S. Dresselhaus, G. Dresselhaus: In *Light Scattering in Solids III*, ed. by M. Cardona, G. Güntherodt, Topics Appl. Phys., Vol.51 (Springer, Berlin, Heidelberg 1982) p.3
5.8 R. Al-Jishi, G. Dresselhaus: Phys. Rev. B **26**, 4514 (1982)
5.9 D.S. Knight, W.B. White: J. Mater. Res. **4**, 385 (1989)
5.10 S.A. Solin, A.K. Ramdas: Phys. Rev. B **1**, 1687 (1970)
5.11 A.R Badzian, P.K. Backmann, T. Hartnett, T. Badzian, R. Messier: In *Amorphous Hydrogenated Carbon Films*, ed. by P. Koidl, P. Oelhafen (Les Editions de Physique, Paris 1987) p.67
5.12 R.J. Nemanich, J.T. Glass, G. Lucovsky, R.E. Shroder: J. Vac. Sci. Technol. A **6**, 1783 (1988)

5.13 J. Wagner, M. Ramsteiner, C. Wild, P. Koidl: Phys. Rev. B **40**, 1817 (1989)
5.14 J.C. Angus, C.C. Hayman: Science **241**, 877 (1988)
5.15 J.C. Angus, C.C. Hayman: Science **241**, 913 (1988)
5.16 J.A. Woollam, H. Chang, V. Natarajan: Appl. Phys. Commun. **5**, 263 (1985)
5.17 M. Ramsteiner, J. Wagner, Ch. Wild, P. Koidl: J. Appl. Phys. **62**, 729 (1987)
5.18 I. Sela, M. Adel, R. Beserman: J. Appl. Phys. **68**, 70 (1990)
5.19 M. Ramsteiner, J. Wagner: Appl. Phys. Lett. **51**, 1355 (1987)
5.20 B. Dischler, A. Bubenzer, P. Koidl, G. Brandt, O.F. Schirmer: In Proc. Symp. Laser Induced Damage in Optical Materials. NBS-SP-669, Boulder, CO (1982)
5.21 B.S. Elman, L. McNeil, C. Nicolini, T.C. Chieu, M.S. Dresselhaus, G. Dresselhaus: Phys. Rev. B **28**, 7201 (1983)
5.22 N. Savvides: J. Appl. Phys. **58**, 518 (1985)
5.23 N. Savvides: J. Appl. Phys. **59**, 4133 (1986)
5.24 L. Salamanca-Riba, G. Braunstein, M.S. Dresselhaus, J.M. Gibson, M. Endo: Nucl. Instr. Meth. Phys. Res. **B7,8**, 487 (1985)
5.25 S. Prawer, D. McCulloch, C.J. Rossouw, S. Glanvill: In *Surface Modification Techniques III*, ed. by T.S. Sudarshan, D.G. Bhat (The Metallurgical Society, Inc., Warrendale, PA 1990)
5.26 P.G. Lurie, J.M. Wilson: Surf. Sci. **65**, 476 (1977)
5.27 T.W. Haas, J.T. Grant, G.J. Dooley: J. Appl. Phys. **43**, 853 (1972)
5.28 B.E. Williams, J.T. Glass: J. Mater. Res. **4**, 373 (1989)
5.29 V.V. Khvostov, M.B. Guseva, V.G. Babaev, O.Yu. Rylova: Solid State Commun. **55**, 443 (1985)
5.30 A. Hoffman, P.J.K. Paterson, S. Prawer: Nucl. Instr. Meth Phys. Res. **52**, 63 (1990)
5.31 A. Hoffman, S. Prawer: In *Surface Chemistry and Beam-Solid Interactions*, ed. by H. Atwater, F.A. Houle, D. Lowndes. MRS Proc. **201**, 619 (1991)
5.32 A. Hoffman, G. Nyberg, S. Prawer: J. Phys. Condensed Matter **2**, 8099 (1990)
5.33 A. Hoffman, S. Prawer, M. Folman: Phys. Rev. B **44**, 4640 (1991)
5.34 K. Ueno, T. Kumihashi, K. Saiki, A. Koma: Jpn. J. Appl. Phys. Lett. **27**, L759 (1988)
5.35 W.P. Hoffman, W.C. Hurley, T.W. Owens, H.T. Phan: Ext. Abstracts of Carbone 90 (Paris 1990)
5.36 A. Hoffman, S. Prawer: Appl. Phys. Lett. **58**, 361 (1991)
5.37 J.A. Golovchenko: Science **232**, 48 (1986)
5.38 N.D. Lang: Phys. Rev. B **34**, 5947 (1986)
5.39 J. Tersoff, D.R. Hamann: Phys. Rev. B **31**, 805 (1985)
5.40 J. Tersoff: Phys. Rev. B **40**, 11990 (1989)
5.41 J. Schneir, R. Sonnenfeldt, P.K. Hansma, J. Tersoff: Phys. Rev. B **34**, 4979 (1986)
5.42 V.B. Elings, F. Wudl: J. Vac. Sci. Technol. A **6**, 412 (1988)
5.43 C.M. Mate, R. Erlandsson, G.M. McClelland, S. Chiang: Surf. Sci. **208**, 473 (1989)
5.44 A. Hoffman, P.J.K. Paterson, S. Prawer: Nucl. Instr. Meth. Phys. Res. B **51**, 226 (1990)
5.45 W.P. Hoffman, H.T. Phan, T.W. Owens, W.C. Hurley: Extended Abstracts, 19th Biennial Conf. on Carbon (1989) p.238
5.46 W.P. Hoffman, V.B. Elings, J.A. Gurley: Carbon **26**, 754 (1988)
5.47 R. Wiesendanger, R. Schlogl, H.J. Güntherodt: Extended Abstracts, Carbon'86 (Baden-Baden 1986) p.207
5.48 S.P. Kelty, C.M. Lieber: Phys. Rev. B **40**, 5856 (1989)
5.49 D. Anselmetti, R. Wiesendanger, H.J. Güntherodt: Phys. Rev. B **39**, 135 (1989)
5.50 R. Coratger, A. Claverie, F. Ajustron, J. Beauvillain: Surf. Sci. **227**, 7 (1990)
5.51 D. Tomanek, S.G. Louie, H.J. Mamin, D.W. Abraham, R.E. Thomson, E. Ganz, J. Clark: Phys. Rev. B **35**, 7790 (1987)

5.52 H.J. Mamin, E. Ganz, D.W. Abraham, R.E. Thomson, J. Clark: Phys. Rev. B **34**, 9015 (1988)
5.53 C.R. Leavens, G.C. Aers: Phys. Rev. B **38**, 7357 (1988)
5.54 J. Tersoff: Phys. Rev. Lett. **57**, 440 (1986)
5.55 C.H. Olk, J. Heremans, M.S. Dresselhaus, J.S. Speck, J.T. Nicholls: Phys. Rev. B **42**, 7524 (1990)
5.56 G. Wagoner: Phys. Rev. **118**, 647 (1960)
5.57 L.S. Singer, G. Wagoner: J. Chem. Phys. **37**, 1812 (1962)
5.58 Y. Kazumata, S. Yugo, T. Kimura, Y. Sato, Y. Nakano: J. Phys. Chem. Solids **47**, 617 (1986)
5.59 Y. Kazumata, S. Yugo, T. Kimura, Y. Nakano: J. Phys. Chem. Solids **47**, 633 (1986)
5.60 J.W. McClure, Y. Yafet: In *Proc. 5th Conf. on Carbon*, ed. by S. Mrozowski, M.L. Studebaker, P.L. Walker Jr. (Pergamon, New York 1962) p.22
5.61 Y. Kazumata, S. Yugo: J. Phys. Soc. Jpn. **51**, 3755 (1982)
5.62 F.J. Dyson: Phys. Rev. **98**, 349 (1955)
5.63 M.E. Adel, R. Kalish, S. Prawer: J. Appl. Phys. **62**, 4096 (1987)
5.64 M.S. Dresselhaus, G. Dresselhaus, K. Sugihara, I.L. Spain, H.A. Goldberg: *Graphite Fibers and Filaments*, Springer Ser. Mater. Sci., Vol.5 (Springer, Berlin, Heidelberg 1988)
5.65 N.E.W. Hartley, M.J. Poole: Mater. Sci. **8**, 900 (1973)
5.66 N.E.W. Hartley, G. Dearnaley, J.F. Turner, J. Saunders: In *Applications of Ion Beams to Metals*, ed. by S.T. Picraux, E.P. Nisse, F.L. Vook (Plenum, New York 1974) p.125

Chapter 6

6.1 B.S. Elman, G. Braunstein, M.S. Dresselhaus, G. Dresselhaus, T. Venkatesan, J.M. Gibson: Phys. Rev. B **29**, 4703 (1984)
6.2 B.S. Elman, M. Shayegan, M.S. Dresselhaus, H. Mazurek, G. Dresselhaus: Phys. Rev. B **25**, 4142 (1982)
6.3 T. Venkatesan, B.S. Elman, G. Braunstein, M.S. Dresselhaus, G. Dresselhaus: J. Appl. Phys. **56**, 3232 (1984)
6.4 B.S. Elman, M.S. Dresselhaus, G. Dresselhaus, E.W. Maby, H. Mazurek: Phys. Rev. B **24**, 1027 (1981)
6.5 P. Lespade, R. Al-Jishi, M.S. Dresselhaus: Carbon **20**, 427 (1982)
6.6 R.O. Dillon, J.A. Woollam, V. Katkanant: Phys. Rev. B **29**, 3482 (1984)
6.7 J. Wagner, M. Ramsteiner, C. Wild, P. Koidl: Phys. Rev. B **40**, 1817 (1989)
6.8 M. Ramsteiner, J. Wagner: Appl. Phys. Lett. **51**, 1355 (1987)
6.9 M. Yoshikawa, G. Katagiri, H. Ishida, A. Ishitani, T. Akamatsu: Solid State Commun. **66**, 1177 (1988)
6.10 M. Yoshikawa, G. Katagiri, H. Ishida, A. Ishitani, T. Akamatsu: Appl. Phys. Lett. **52**, 1639 (1988)
6.11 M. Yoshikawa, G. Katagiri, H. Ishida, A. Ishitani, T. Akamatsu: J. Appl. Phys. **64**, 6464 (1988)
6.12 R.P. Vidano, D.B. Fishbach, L.J. Willis, T.M. Loehr: Solid State Commun. **39**, 341 (1981)
6.13 R. Coratger, A. Claverie, F. Ajustron, J. Beauvillain: Surf. Sci. **227**, 7 (1990)
6.14 L. Porte, M. Phaner, C.H. de Villeneuve, N. Moncoffre, J. Tousset: Nucl. Instr. Meth. Phys. Res. B **44**, 116 (1989)
6.15 L. Porte, C.H. de Villeneuve, M. Phaner: J. Vac. Sci. Tech. **9**, 1064 (1991)
6.16 H.A. Mizes, S. Park, W.A. Harrison: Phys. Rev. B **36**, 4491 (1987)
6.17 A. Hoffman, P.J.K. Paterson, S. Prawer: Nucl. Instr. Meth. Phys. Res. B **51**, 226 (1990)

6.18 A. Hoffman, S. Prawer: In *Surface Chemistry and Beam-Solid Interactions*, ed. by H. Atwater, F.A. Houle, D. Lowndes. MRS Proc. **201**, 619 (1991)
6.19 Y. Kazumata, S. Yugo: J. Phys. Soc. Jpn. **51**, 3755 (1982)
6.20 Y. Kazumata, S. Yugo, T. Kimura, Y. Sato, Y. Nakano: J. Phys. Chem. Solids **47**, 617 (1986)
6.21 Y. Kazumata, S. Yugo, T. Kimura, Y. Nakano: J. Phys. Chem. Solids **47**, 633 (1986)
6.22 L. Salamanca-Riba, G. Braunstein, M.S. Dresselhaus, J.M. Gibson, M. Endo: Nucl. Instr. Meth. Phys. Res. B **7,8**, 487 (1985)
6.23 M. Endo, L. Salamanca-Riba, G. Dresselhaus, J.M. Gibson: J. Chimie Physique **81**, 803 (1984)
6.24 B.T. Kelly: *Physics of Graphite* (Applied Science, London 1981)
6.25 S. Yugo, T. Kimura, Y. Kazumata: Carbon **23**, 147 (1985)
6.26 D.E. Soule, C.W. Nezbeda: J. Appl. Phys. **39**, 5122 (1968)
6.27 J.E. Brocklehurst: In *Chemistry and Physics of Carbon*, ed. by P.L. Walker Jr, P.A. Thrower (Dekker, New York 1976) p.145
6.28 D. Schroyen, M. Bruggeman, I. Dezsi, G. Langouche: Nucl. Instr. Meth. Phys. Res. B **15**, 341 (1986)
6.29 B. Söder, J. Roth, W. Möller: Phys. Rev. B **37**, 815 (1988)
6.30 B.S. Elman, H. Mazurek, M.S. Dresselhaus, G. Dresselhaus: In *Metastable Materials Formation by Ion Implantation*, ed. by S.T. Picraux, W.J. Choyke (Elsevier, New York 1982) p.425
6.31 A. Oberlin: Carbon **22**, 521 (1984)
6.32 B.S. Elman, G. Braunstein, M.S. Dresselhaus, G. Dresselhaus, T. Venkatesan, B. Wilkens: J. Appl. Phys. B **56**, 2114 (1984)
6.33 Y. Kazumata: J. Phys. Chem. Solids **44**, 1025 (1983)
6.34 F.F. Morehead Jr., B.L. Crowder: Rad. Eff. **6**, 27 (1970)
6.35 M. Endo, T.C. Chieu, G. Timp, M.S. Dresselhaus, B.S. Elman: Phys. Rev. B **28**, 6982 (1983)
6.36 B.S. Elman, G. Braunstein, M.S. Dresselhaus, T. Venkatesan: Nucl. Instr. Meth. Phys. Res. B **7** & **8**, 493 (1985)
6.37 B.T. Kelly, J.E. Brocklehurst: Carbon **9**, 783 (1971)
6.38 W.S. Reynolds: In *Physical Properties of Graphite* (Elsevier, New York 1968)
6.39 B.S. Elman, L. McNeil, C. Nicolini, T.C. Chieu, M.S. Dresselhaus, G. Dresselhaus: Phys. Rev. B **28**, 7201 (1983)
6.40 M.J. Kenny, J.T.A. Pollock, L.S. Wielunski: Nucl. Instr. Meth. Phys. Res. B **39**, 704 (1989)
6.41 J.W. McClure: In *Proc. Int'l Conf. on Semimetals and Narrow Gap Semiconductors*, ed. by D.L. Carter, R.T. Bate (Pergamon, New York 1971) p.127
6.42 S. Yugo, T. Kimura: Jpn. J. Appl. Phys. **22**, 1738 (1983)
6.43 J.K. Hirvonen: In *Ion Implantation and Ion Beam Processing of Materials*, ed. by G.K. Huber, O.W. Holland, C.R. Clayton, C.W. White (North-Holland, New York 1984) p.621
6.44 J.P. Hirvonen, D. Stone, M. Nastasi, S.P. Hannula: Scr. Metall. **20**, 649 (1986)
6.45 R. Taylor, R.G. Brown, K. Gilchrist, E. Hall, A.T. Hodds, B.T. Kelly, F. Morris: Carbon **5**, 519 (1967)
6.46 Y. Matsuhisa, M. Washiyama, T. Hiramatsu, H. Fujino, G. Katagiri: Extended Abstracts, 20th Biennial Conf. on Carbon (Santa Barbara, CA 1991) p.226
6.47 N.C. Koon, P. Pehrsson, D. Weber, A.I. Schindler: J. Appl. Phys. **55**, 2497 (1984)
6.48 A. Lusnikov, I. Ohana, M.S. Dresselhaus, S.P. Withrow: J. Appl. Phys. **63**, 195 (1987)
6.49 H. Menjo, B.S. Elman, G. Braunstein, M.S. Dresselhaus: J. Chimie Physique **81**, 835 (1984)
6.50 K. Ashida, K. Kanamori, K. Watanabe: J. Vac. Sci. Tech. A **6**, 2232 (1988)

6.51 J. Roth: J. Nucl. Mater. **145-147**, 87 (1987)
6.52 W. Möller, J. Roth: In *Physics of Plasma-Wall Interactions in Controlled Fusion*, ed. by D.E. Post, R. Behrisch (Plenum, New York 1986) p.439
6.53 K.L. Wilson, W.L. HsuW: J. Nucl. Mater. **145-147**, 121 (1987)
6.54 E. Vietzke, V. Philipps: Nucl. Instr. Meth. Phys. Res. B **23**, 449 (1987)
6.55 W.R. Wampler, D.K. Brice, C.W. Magee: J. Nucl. Mater. **102**, 304 (1981)
6.56 W. Möller, B.M.U. Scherzer: Appl. Phys. Lett. **50**, 1870 (1987)
6.57 B.M.U. Scherzer, J. Wang, W. Möller: J. Nucl. Mater. **162-164**, 1013 (1989)
6.58 B.M.U. Scherzer, M. Wielunski, W. Möller, A. Turos, J. Roth: Nucl. Instr. Meth. Phys. Res. B **33**, 714 (1988)
6.59 B.M.U. Scherzer, J. Wang, W.,Möller: Nucl. Instr. Meth. Phys. Res. B **45**, 54 (1990)
6.60 W. Möller: J. Nucl. Mater. **162-164**, 138 (1989)
6.61 J. Winter: J. Nucl. Mater. **145-147**, 131 (1987)
6.62 J. Roth, B.M.U. Scherzer, R.S. Blewer, D.K. Brice, S.T. Picraux, W.R Wampler: J. Nucl. Mater. **93-94**, 601 (1980)
6.63 C.K. Chen, B.M.U. Scherzer, W. Eckstein: Appl. Phys. A **33**, 265 (1984)
6.64 B.M.U. Scherzer: Nucl. Instr. Meth. Phys. Res. B **45**, 57 (1990)
6.65 B.M.U. Scherzer: J. Nucl. Mater. **168**, 121 (1989)
6.66 D.K. Brice, B.L. Dole, W.R. Wampler: J. Nucl. Mater. **111-112**, 598 (1982)

Chapter 7

7.1 T. Venkatesan, L. Calcagno, B.S. Elman, G. Foti: In *Ion Beam Modification of Insulators*, ed. by P. Mazzoldi, G.W. Arnold (Elsevier, New York 1987) Vol.2, p.301
7.2 J.T.A. Pollock, M.J. Kenny, L.S. Wielunski, M.D. Scott: NATO Workshop (1989)
7.3 D.F. Baker, R.H. Bragg: Phys. Rev. B **28**, 2219 (1983)
7.4 S. Prawer, F. Ninio, I. Blanchonette: J. Appl. Phys. **68**, 2361 (1990)
7.5 D. McCulloch, S. Prawer: In *Beam-Solid Interactions: Physical Phenomena*, ed. by J.A. Knapp, P. Bϕrgesen, R.A. Zuhr. MRS Proc. **157**, 825 (1989)
7.6 M. Iwaki, K. Takahashi, K. Yoshida, Y. Okabe: Nucl. Instr. Meth. Phys. Res. B **39**, 700 (1989)
7.7 S. Prawer, D. McCulloch, C.J. Rossouw, S. Glanvill: In *Surface Modification Techniques III*, ed. by T.S. Sudarshan, D.G. Bhat (The Metallurgical Society, Inc., Warrendale, PA 1990)
7.8 A.V. Baranov, A.N. Bekhterev, Y.S. Bobovich, V.I. Petrov: Opt. Spectrosc. USSR **62**, 612 (1987)
7.9 M.S. Dresselhaus, G. Dresselhaus: In *Light Scattering in Solids III*, ed. by M. Cardona, G. Güntherodt, Topics Appl. Phys., Vol.51 (Springer, Berlin, Heidelberg 1982) p.3
7.10 R. Al-Jishi, G. Dresselhaus: Phys. Rev. B **26**, 4514 (1982)
7.11 G.M. Jenkins, K. Kawamura: In *Polymeric Carbons - Carbon Fibre, Glass and Char* (Cambridge Univ. Press, London 1976)
7.12 Y. Shiraishi: In *Introduction to Carbon Materials* (Carbon Society of Japan, Tokyo 1984) pp.29-40
7.13 Y. Kaburagi, S. Yasuda, Y. Hishiyama: Extended Abstracts, 18th Biennial Conf. on Carbon (Worcester, MA 1987) p.476
7.14 S. Prawer, C.J. Rossouw: J. Appl. Phys. **63**, 4435 (1988)
7.15 Y. Bar-Yam, T.D. Moustakas: Nature **342**, 786 (1989)
7.16 D. McCulloch, S. Prawer, D. Sengupta: In *Surface Chemistry and Beam-Solid Interactions*, ed. by H. Atwater, F.A. Houle, D. Lowndes. MRS Proc. **201**, 301 (1990)

7.17 M.J. Kenny, J.T.A. Pollock, L.S. Wielunski: Nucl. Instr. Meth. Phys. Res. **B39**, 704 (1989)
7.18 M. Iwaki, K. Takahashi, A. Sekiguchi: J. Mater. Res. **5**, 2562 (1990)
7.19 K. Yoshida, K. Takahashi, K. Okuno, G. Katagiri, M. Iwaki, A. Ishitani: Appl. Phys. Lett. **52**, 1046 (1988)
7.20 J.P. Hirvonen, D. Stone, M. Nastasi, S.P. Hannula: Scr. Metall. **20**, 649 (1986)
7.21 J.T.A. Pollock, R.A. Clissold, M. Farrelly: In *Materials Modification and Growth Using Ion Beams*, ed. by U. Gibson, A.E. White, P.P. Pronko. MRS Proc. **93**, 317 (1987)
7.22 M. Farrelly, J.T.A. Pollock: Mater. Forum **10** 198 (1987)
7.23 M.S. Dresselhaus, G. Dresselhaus, K. Sugihara, I.L. Spain, H.A. Goldberg: *Graphite Fibers and Filaments*, Springer Ser. Mater. Sci., Vol.5 (Springer, Berlin, Heidelberg 1988)
7.24 L. Salamanca-Riba, G. Braunstein, M.S. Dresselhaus, J.M. Gibson, M. Endo: Nucl. Instr. Meth. Phys. Res. B **7/8**, 487 (1985)
7.25 M. Endo, L. Salamanca-Riba, G. Dresselhaus, J.M. Gibson: J. Chimie Physique **81**, 803 (1984)
7.26 M. Endo, T.C. Chieu, G. Timp, M.S. Dresselhaus, B.S. Elman: Phys. Rev. B **28**, 6982 (1983)
7.27 S. Yugo, T. Misaki, T. Kimura: Tanso **108**, 2 (1981) (in Japanese)
7.28 O.C. de Hodgins, M.S. Dresselhaus, D. Uhlmann, C.F. Aliotta: In *Electronic Packaging Materials Science*, ed. by E.A. Giess, K.N. Tu, D.R. Uhlmann (Elsevier, New York 1985) pp.111-116
7.29 Y. Matsuhisa, M. Washiyama, T. Hiramatsu, H. Fujino, G. Katagiri: Extended Abstracts, 20th Biennial Conf. on Carbon (Santa Barbara, CA 1991) p.226
7.30 J.J. Hauser: Solid State Commun. **17**, 1577 (1975)
7.31 T. Venkatesan, R. Levi, T.C. Banwell, T. Tombrello, M. Nicolet, R. Hamm, A.E. Meixner: In *Ion Beam Processes in Advanced Electronic Materials and Device Technology*, ed. by B.B. Appleton, F.H. Eisen, T.W. Sigmon (MRS, Pittsburgh 1985) p.189
7.32 T. Venkatesan, T., R.C. Dynes, B. Wilkens, A.E. White, J.M. Gibson, R. Hamm: Nucl. Instr. Meth. Phys. Res. **B1**, 599 (1984)
7.33 B. Wasserman, G. Braunstein, M.S. Dresselhaus, G.E. Wnek: In *MRS Symp. on Ion Implantation and Ion Beam Processing of Materials*, ed. by G.K. Hubler, O.W. Holland, C.R. Clayton, C.W. White (Elsevier, New York 1984) p.423
7.34 T. Wada, A. Takeno, M. Iwake, H. Sasabe, Y. Kobayashi: J. Chem. Soc. Chem. Commun. **17**, 1194 (1985)

Chapter 8

8.1 S. Prawer, R. Kalish: Unpublished
8.2 V.S. Vavilov, V.V. Krasnopevtsev, Y.V. Miljutin, A.E. Gorodetsky, A.P. Zakharov: Rad. Eff. **22**, 141 (1974)
8.3 A. Hoffman, P.J.K. Paterson, S. Prawer: Nucl. Instr. Meth. Phys. Res. B **51**, 226 (1990)
8.4 S. Prawer, F. Ninio, I. Blanchonette: J. Appl. Phys. **68**, 2361 (1990)
8.5 R. Kalish, T. Bernstein, B. Shapiro, A. Talmi: Rad. Eff. **52**, 153 (1980)
8.6 I. Miyamoto, K. Nishimura, K. Kawata, H. Kawarada, S. Shimada: Nucl. Instr. Meth. Phys. Res. **B39**, 689 (1989)
8.7 J.J. Hauser, J.R. Patel: Solid State Commun. **18**, 789 (1976)
8.8 J.J. Hauser, J.R. Patel, J.N. Rogers: Appl. Phys. Lett. **30**, 129 (1977)
8.9 J.F. Prins: Phys. Rev. B **31**, 2472 (1985)
8.10 S. Sato, M. Iwaki: Nucl. Instr. Meth. Phys. Res. B **32**, 145 (1988)
8.11 J.F. Prins: Rad. Eff. Lett. **76**, 79 (1983)

8.12 E.W. Maby, C.W. Magee, J.H. Morewood: Appl. Phys. Lett. **39**, 157 (1981)
8.13 J.F. Prins, T.E. Derry, J.P.F. Sellschop: Phys. Rev. **B34**, 8870 (1986)
8.14 J.F. Prins, T.E. Derry, J.P.F. Sellschop: Nucl. Instr. Meth. Phys. Res. B **18**, 261 (1987)
8.15 R.A. Spits, T.E. Derry, J.F. Prins, J.P.F. Sellschop: Nucl. Instr. Meth. Phys. Res. **B51**, 63 (1990)
8.16 T.E. Derry, J.F. Prins, C.C.P. Madiba, J. Ennis, R.A. Spits, J.P.F. Sellschop: Nucl. Instr. Meth. Phys. Res. B **35**, 431 (1988)
8.17 J. Bernholc, A. Antonelli, T.M. Del Sole, Y. Bar-Yam, S.T. Pantelides: Phys. Rev. Lett. **61**, 2689 (1988)
8.18 J.P.F. Sellschop: In *Properties of Diamond*, ed. by J.E. Field (Academic, London 1979) p.107
8.19 C.D. Clark, P.J. Kemmey, E.W.J. Mitchell: Discuss. Faraday Soc. **31**, 96 (1961)
8.20 J.C. Bourgoin, B. Massarani: Phys. Rev. B **14**, 3690 (1976)
8.21 M. Shaanan: Priv. communication (1990)
8.22 S. Prawer, A. Hoffman, R. Kalish: Unpublished
8.23 G. Braunstein, A. Talmi, R. Kalish, T. Bernstein, R. Beserman: Rad. Eff. **48**, 139 (1980)
8.24 T.E. Derry: Preprint of Hamilton Conf. See NIMB proc. Ion Beam Analysis or Modification Conf. in Hamilton (1987)
8.25 T.E. Derry, R.A. Spits, J.P.F. Sellschop: In *Properties and Applications of SiC, Natural and Synthetic Diamond and Related Materials*, ed. by A.A. Gippius, R. Helbig, J.P.F. Sellschop (Les Editions de Physique, Paris 1991)
8.26 G. Braunstein, R. Kalish: J. Appl. Phys. **54**, 2106 (1983)
8.27 J.F. Morhange, R. Beserman, J.C. Bourgoin: Jpn. J. Appl. Phys. **14**, 544 (1975)
8.28 M. Teicher, R. Beserman: J. Appl. Phys. **53**, 1467 (1982)
8.29 A. Hoffman, S. Prawer: In *Surface Chemistry and Beam-Solid Interactions*, ed. by H. Atwater, F.A. Houle, D. Lowndes. MRS Proc. **201**, 619 (1991)
8.30 A. Hoffman, S. Prawer: Appl. Phys. Lett. **58**, 361 (1991)
8.31 J.F. Prins: Phys. Rev. B **38**, 5576 (1988)
8.32 G. Braunstein, T. Bernstein, U. Carsenty, R. Kalish: J. Appl. Phys. **50**, 5731 (1979)
8.33 B. Liu, G.S. Sandhu, N.R. Parikh, M.L. Swanson: Nucl. Instr. and Meth. B **45**, 420 (1990)
8.34 Y.H. Lee, P.R. Brousious, J.W. Corbett: Phys. Stat. Sol. A **50**, 237 (1978)
8.35 M.E. Adel, R. Kalish, V. Richter: J. Mater. Res. **1**, 503 (1986)
8.36 R.G. Elliman, M.C. Ridgeway, J.S. Williams: In *Beam-Solid Interactions: Physical Phenomena*, ed. by J.A. Knapp, P. Bϕrgesen, R.A. Zuhr. MRS Proc. **157**, 105 (1989)
8.37 V.S. Vavilov: Phys. Stat. Sol. A **31**, 11 (1975)
8.38 M.W. Geis, N.N. Efremow, D.D. Rathman: J. Vac. Sci. Tech. A **6** 1953 (1988)
8.39 M.W. Geis: In *Diamond, Boron Nitride, Silicon Carbide and Related Wide Bandgap Semiconductors*, ed. by J.T. Glass, R.F. Messier, N. Fujimori. MRS Proc. **162**, 15 (MRS, Pittsburgh 1990)
8.40 S.A. Kajihara, A. Antonelli, J. Bernholc, R. Car: Phys. Rev. Lett. **66**, 2010 (1991)
8.41 J. Prins: In *Properties and Applications of SiC, Natural and Synthetic Diamond and Related Materials*, ed. by A.A. Gippius, R. Helbig, J.P.F. Sellschop (Les Editions de Physique, Paris 1991)
8.42 G. Braunstein, R. Kalish: Nucl. Instr. Meth. Phys. Res. **182–183**, 691 (1981)
8.43 V.S. Vavilov: Rad. Eff. **37**, 229 (1978)
8.44 B.M. Vul: Jpn. J. Appl. Phys. **43**, 183 (1974)
8.45 G.S. Sandhu, W.K. Chu, M.L. Swanson, J.F. Prins: Proc. SPIE **969**, 37 (1985)
8.46 L.A. Davidson, S. Chou, J.F. Gibbons, W.S. Johnson: Rad. Eff. **7**, 35 (1971)

8.47 V.S. Vavilov, E.A. Konorova, E.B. Stepanova, E.M. Turkhan: Fiz. Tekh. Poluprovodn **13**, 1033 (1979)
8.48 V.S. Vavilov, E.A. Konorova, E.B. Stepanova, E.M. Turkhan: Fiz. Tekh. Poluprovodn **13**, 1083 (1979)
8.49 G.S. Sandhu, M.L. Swanson, W.K. Chu: In *Processing and Characterization of Materials Using Ion Beams*, ed. by L.E. Rehn, J. Greene, F.A. Smidt. MRS Proc. **128**, 707 (MRS Press, Pittsburgh 1989)
8.50 R. Kalish, M. Deicher, E. Recknagel, T. Wichert: J. Appl. Phys. **50**, 6870 (1979)
8.51 H. Appel, J. Raudies, W.G. Thies, A. Hanser, J.P.F. Sellschop: Hyperfine Interactions **10**, 735 (1981)
8.52 J.H. Raudies, H. Appel, G.M. Then, W.G. Thies, K. Frietag, J.P.F. Sellschop, M.C. Stemmet: Hyperfine Interactions **15-16**, 487 (1983)
8.53 S. Connell, K. Bharuth-Ram, J.P.F. Sellschop, M.C. Stemmet, H. Appel: Nucl. Instr. Meth. Phys. Res. **B35**, 423 (1988)
8.54 W. Werwoerd: Nucl. Instr. Meth. Phys. Res. **B35**, 509 (1988)
8.55 M. van Rossum, J. de Bruyn, G. Langouche, M. de Potter, R. Coussement: Phys. Lett. **73A**, 127 (1979)
8.56 M. van Rossum, G. Langouche, J. de Bruyn, M. de Potter, R. Coussement: Nucl. Instr. Meth. Phys. Res. **182--183**, 407 (1981)
8.57 M. de Potter, G. Langouche: Hyperfine Interactions **15**, 479 (1983)
8.58 M. de Potter, G. Langouche: Z. Physik B **53**, 89 (1983)
8.59 J.A. Sawicki, B.D. Sawicka: Nucl. Instr. Meth. Phys. Res. **194**, 465 (1982)
8.60 J.A. Sawicki, B.D. Sawicka, H. de Waard: Hyperfine Interactions **15**, 483 (1983)
8.61 G. Langouche: Hyperfine Interactions **29**, 1283 (1986)
8.62 G.H. Glover: Solid State Electron. **16**, 973 (1973)
8.63 J.F. Prins: Appl. Phys. Lett. **41**, 950 (1982)
8.64 J.P. Akimchenko, K.V. Kisseleva, V.V. Krasnopevtsev, Y.V. Milyutin, A.G. Touryansky, V.S. Vavilov: Rad. Eff. **33**, 75 (1977)
8.65 J.P. Akimchenko, V.S. Vavilov, V.V. Krasnopevtsev, Y.V. Milyutin, J. Gyulai, G. Mezey, T. Nagy: Inst. Phys. Conf. Ser. (London) **31**, 354 (1977)
8.66 V.V. Krasnopevtsev, Y.V. Milyutin, V.S. Vavilov, A.E. Gorodetskii, A.N. Khodan, A.P. Zakharov: In *Proc. 5th Int'l Conf. Ion Impl. Semicond.*, ed. by F. Chernow, J.A. Borders, D.K. Brice (Plenum, New York 1977) p.295
8.67 R.S. Nelson, J.A. Hudson, D.J. Mazey, R.C. Piller: Proc. Roy. Soc. London A **386**, 211 (1983)
8.68 R.S. Nelson: British Patent No. 1476313 (1977)
8.69 T.E. Derry, J.P.F. Sellschop: Nucl. Instr. Meth. Phys. Res. **191**, 23 (1981)
8.70 N.E.W. Hartley, M.J. Poole: Mater. Sci. **8**, 900 (1973)
8.71 N.E.W. Hartley: In *Metastable Materials Formation by Ion Implantation*, ed. by S.T. Picraux, W.J. Choyke (North-Holland, Amsterdam 1982) p.295
8.72 W.L. Lin: Beijing Shifan Daxue Xeubau (Bulletin of Beijing Teachers College) **3**, 39 (1984)

Chapter 9

9.1 J. Aisenberg, R. Chabot: J. Appl. Phys. **42**, 2953 (1971)
9.2 S. Prawer, A. Hoffman, R. Kalish: Appl. Phys. Lett. **57**, 2187 (1990)
9.3 M.E. Adel: Ion and Laser Beam Induced Modifications of Diamond-like Carbon Coatings. Dissertation, Israel Institute of Technology, Technion, Haifa (1989)
9.4 R. Kalish, E. Adel: In *Properties and Preparation of Amorphous Carbon Films*, ed. by J.J. Pouch, S.A. Alterowitz (Trans Tech., Aedermannsdorf, Switzerland 1990) p.427
9.5 R. Kalish: In *Diamond and Diamond-Like Coatings*, ed. by R.E. Clausing, L.L. Horton, J.C. Angus, P. Koidl (Plenum, New York 1991) p.447

9.6 S. Prawer, R. Kalish, M.E. Adel, V. Richter: J. Appl. Phys. **61**, 4492 (1987)
9.7 D.C. Ingram, D.J. Ehrlich: J. Vac. Sci. Technol. B **4**, 310 (1986)
9.8 D.C. Ingram, A.W. McCormick: Nucl. Instr. Meth. Phys. Res. B **34**, 68 (1988)
9.9 M. Ramsteiner, J. Wagner: Appl. Phys. Lett. **51**, 1355 (1987)
9.10 I. Sela, M. Adel, R. Beserman: J. Appl. Phys. **68**, 70 (1990)
9.11 D.S. Knight, W.B. White: J. Mater. Res. **4**, 385 (1989)
9.12 J. Gonzales-Hernandez, R. Asomoza, A. Reyes-Mena, J. Richards, S.S. Chao, D. Pawlik: J. Vac. Sci. Technol. **A6**, 1798 (1988)
9.13 H. Baumann, T. Rupp, K. Bethge, P. Koidl, C. Wild: In *Amorphous Hydrogenated Carbon Films*, ed. by P. Koidl, P. Oelhafen (Les Editions de Physique, Paris 1987) p.149
9.14 M.E. Adel, O. Amir, R. Kalish, L.C. Feldman: J. Appl. Phys. **66**, 3248 (1989)
9.15 J.P.F. Sellschop, C.C.P. Madiba, H.J. Annegarn: Nucl. Instr. Meth. Phys. Res. **168**, 529 (1980)
9.16 C. Wild, P. Koidl: Appl. Phys. Lett. **51** 19 (1987)
9.17 W. Moeller, P. Bϕrgesen, B.M.U. Scherzer: Nucl. Instr. Meth. Phys. Res. B **19,20**, 826 (1987)
9.18 J.W. Zou, K. Schmidt, K. Reichelt, B. Stritzker: J. Vac. Sci. Tech. A **6**, 3103 (1988)
9.19 F. Fujimoto, M. Tanaka, Y. Iwata, A. Ootuka, K. Komaki, M. Haba, K. Kobayashi: Nucl. Instr. Meth. Phys. Res. B **33**, 792 (1988)
9.20 M.E. Adel, R. Kalish, S. Prawer: J. Appl. Phys. **62**, 4096 (1987)
9.21 S.P. Wong, S. Peng: In *Diamond, Boron Nitride, Silicon Carbide and Related Wide Bandgap Semiconductors*, ed. by J.T. Glass, R.F. Messier, N. Fujimori. MRS Proc. **162**, 383 (MRS, Pittsburgh 1990)
9.22 O. Amir, R. Kalish: In *Diamonds and Related Materials*, ed. by P.K. Bachmann (Elsevier, New York 1992) Paper 7.84
9.23 O. Amir: Doping of a-C:H films by ion-implantation and during growth. M.Sc. Thesis, Technion, Haifa, Israel (1990)
9.24 S.-C. Cheng, D. Fu, Z. Xu, H. Zhang, X.-Z. Pan: Nucl. Instr. Meth. Phys. Res. B **39**, 692 (1989)
9.25 D.R. McKenzie, R.C. McPhedram, N. Savvides, D.J.H. Cockayne: Thin Solid Films **108**, 247 (1983)
9.26 J. Robertson, E.P. O'Reilly: Phys. Rev. B **35**, 2946 (1987)
9.27 S. Prawer, R. Kalish: Unpublished
9.28 S. Yugo, T. Kimura, Y. Kazumata: Carbon **23**, 147 (1985)
9.29 D.E. Soule, C.W. Nezbeda: J. Appl. Phys. **39**, 5122 (1968)

Chapter 10

10.1 R.E. Smalley: The Sciences **31**, 22-30 (March-April 1991)
10.2 J.F. Prins: Phys. Rev. B **38**, 5576 (1988)
10.3 Y. Bar-Yam, T.D. Moustakas: Nature **342**, 786 (1989)

10.4 M.E. Adel, R. Brener, R. Kalish: In *Amorphous Hydrogenated Carbon Films*, ed. by P. Koidl, P. Oelhafen (Les Editions de Physique, Paris 1987) p.335
10.5 S. Prawer, R. Kalish, M.E. Adel: Appl. Phys. Lett. **48**, 1585 (1986)
10.6 S.A. Kajihara, A. Antonelli, J. Bernholc, R. Car: Phys. Rev. Lett. **66**, 2010 (1991)

Subject Index